My Book

This book belongs to

Cover Design by :
Gowri Vemuri

First Edition :
May, 2019

Second Edition :

April, 2021

Author :
Gowri Vemuri

Edited by :
Raksha Pothapragada
Rtvik Pothapragada
Questions: mathknots.help@gmail.com

NOTE : VDOE is neither affiliated nor sponsors or endorses this product.

Dedication

This book is dedicated to:

My Mom, who is my best critic, guide and supporter.

To what I am today, and what I am going to become tomorrow,

is all because of your blessings, unconditional affection and support.

This book is dedicated to the

strongest women of my life ,

my dearest mom

and

to all those moms in this universe.

G.V.

Visit www.a4ace.com

Also available more time based practice tests on subscription

The Virginia Board of Education and Virginia Department of Education (VDOE) have developed the Virginia Assessment Program (VAP) to measure and evaluate students' academic progress in the Standards of Learning (SOLs). The SOLs indicate Virginia's expectations for what students should know and be able to do in the subject areas of reading, writing, mathematics, science, and history/social science.

Students in grades 3-12 to take the Standards of Learning (SOL) assessments each year. Some of the tests are required by all students each year, and others are required only at specific grade levels. Additionally, with the removal of some SOL tests in recent years, the VDOE assigned the responsibility of the creation and administration of alternate, performance - based assessments on local divisions. Student scores from these tests determine a school's and the division's state accreditation and measures progress toward meeting federal targets.

Virginia Standards of Learning (SOL) tests are generally given online unless a student has an identified and documented need to be assessed using paper,pencil format. The test question format is typically multiple choice, and each test contains some technology enhanced items.

GRADE 3-8 STANDARDS OF LEARNING(SOL) TESTS

GRADE 3	GRADE 4	GRADE 5	GRADE 6	GRADE 7	GRADE 8
	VIRGINIA STUDIES				WRITING
MATH	MATH	MATH	MATH	MATH	MATH
READING	READING	READING	READING	READING	READING
		SCIENCE			SCIENCE

NOTE : VDOE is neither affiliated nor sponsors or endorses this product.

END OF COURSE STANDARDS OF LEARNING (SOL) TESTS

GRADE 9	GRADE 10	GRADE 11
ALGEBRA I	GEOMETRY	ALGEBRA II
EARTH SCIENCE	BIOLOGY	CHEMISTRY
WORLD HISTORY I	WORLD HISTORY II	VIRGINIA & US HSTORY
		WORLD GEOGRAPHY
		ENGLISH : READING
		ENGLISH : WRITING

Any Student taking one of the courses listed here is expected to take the corresponding end-of-course SOL test. The grade levels depicted here represent grade level at which students typically participate in these courses.

NOTE : VDOE is neither affiliated nor sponsors or endorses this product.

SOL Test Scoring and Performance Reports:

Standards of Learning assessments in English reading, mathematics, science and history/social science are made up of 35-50 items or questions that measure content knowledge, scientific and mathematical processes, reasoning and critical thinking skills. English writing skills are measured with a two-part assessment that includes multiple-choice items and an essay.

Student performance is graded on a scale of 0-600 with 400 representing the minimum level of acceptable proficiency and 500 representing advanced proficiency. On English reading and mathematics tests, the Board of Education has defined three levels of student achievement: basic, proficient, and advanced, with basic describing progress towards proficiency.

Performance Achievement Levels:

- The achievement levels for grades 3-8 reading and mathematics tests are: *Pass/Advanced*, *Pass/Proficient*, *Fail/Basic*, and *Fail/Below Basic*.

- The achievement levels for science tests, history tests, and End-of-Course (EOC) tests* are: *Pass/Advanced*, *Pass/Proficient*, and *Fail/Does Not Meet*.

- The EOC Writing (2010 SOL) test, EOC Reading (2010 SOL) test, and EOC Algebra II (2009 SOL) test have an achievement level of *Advanced/College Path* in place of the *Pass/Advanced* achievement level.

NOTE : VDOE is neither affiliated nor sponsors or endorses this product.

INDEX

Contents	Page No
Preface	1 - 10
Formulae	13 - 16
Practice Test - 4	17 - 46
Practice Test - 5	47 - 76
Practice Test - 6	77 - 106
Practice Test - 4 Answer keys	107 - 112
Practice Test - 5 Answer keys	113 - 118
Practice Test - 6 Answer keys	119 - 124
Score calculation	125 - 130

FORMULA SHEET

1. Area of a triangle

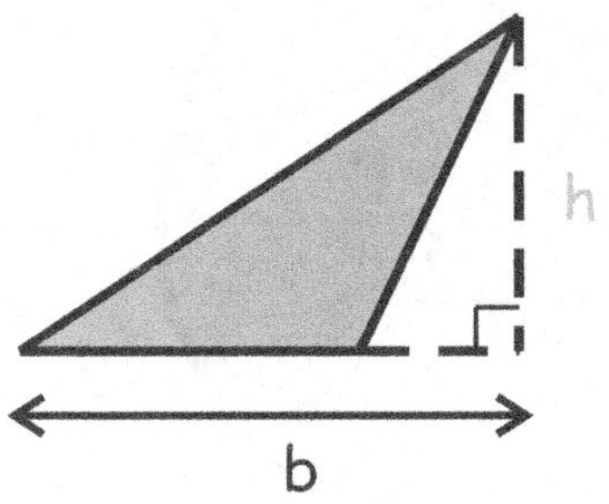

$A = \frac{1}{2} bh$

2. Area of a parellelogram

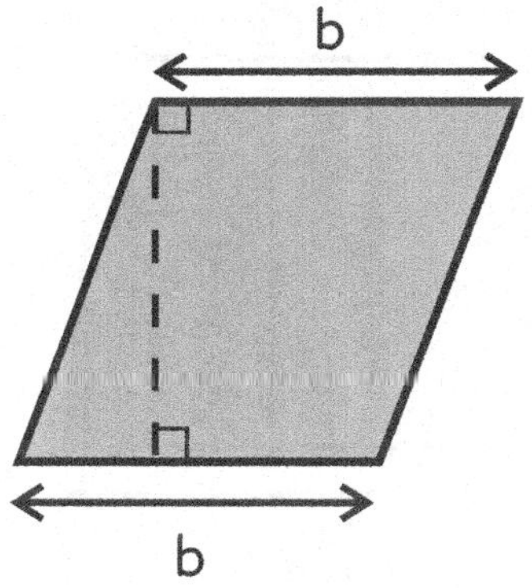

$A = bh$

3. Volume and Surface area of a cuboid

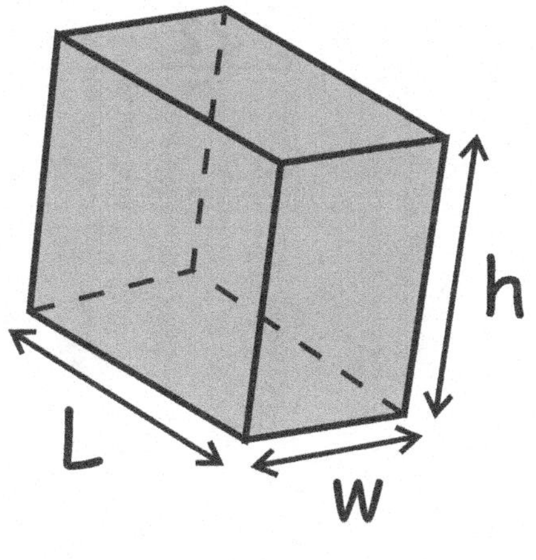

$V = lwh$

$S.A = 2(lw + lh + wh)$

4. Perimeter and Area of a Square

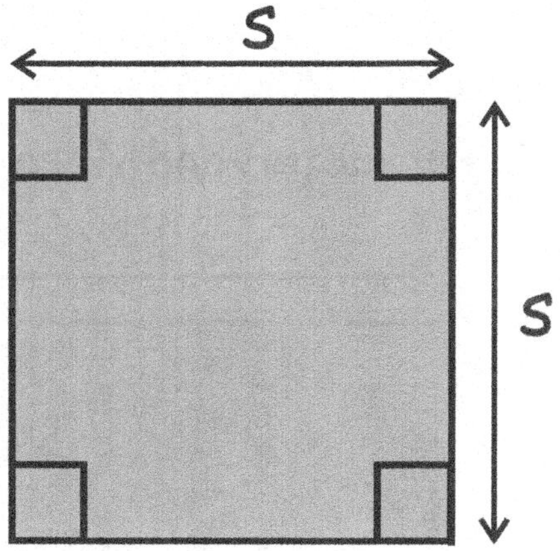

$P = 4s$

$A = s^2$

5. Area of a Trapezium

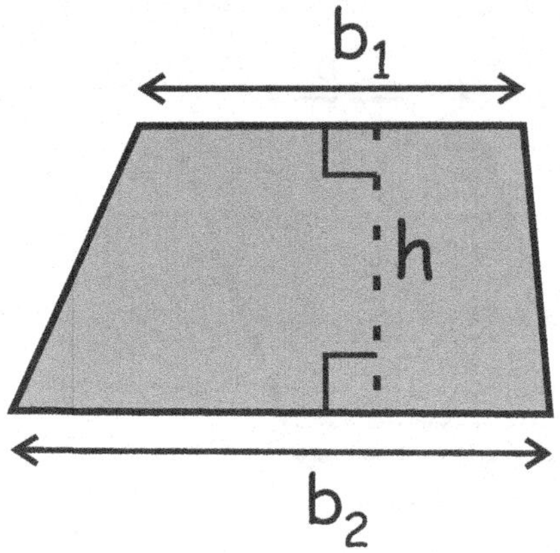

$$A = \frac{1}{2} h(b_1 + b_2)$$

6. Circumference and Area of a Circle

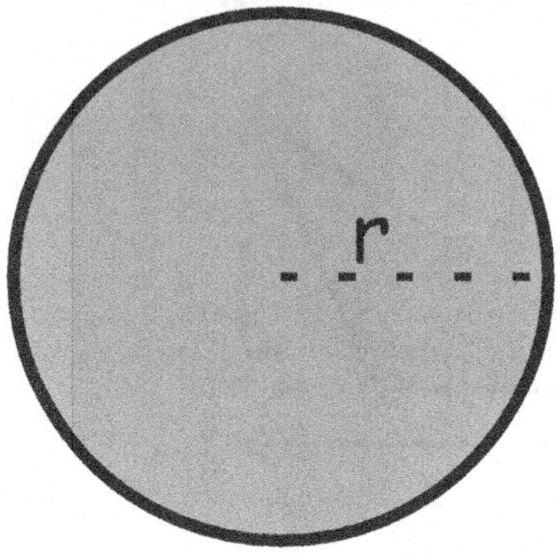

$$c = 2\Pi r$$
$$A = \Pi r^2$$

pi

$$\Pi = 3.14$$
$$\Pi = \frac{22}{7}$$

7. Perimeter and Area of a Rectangle

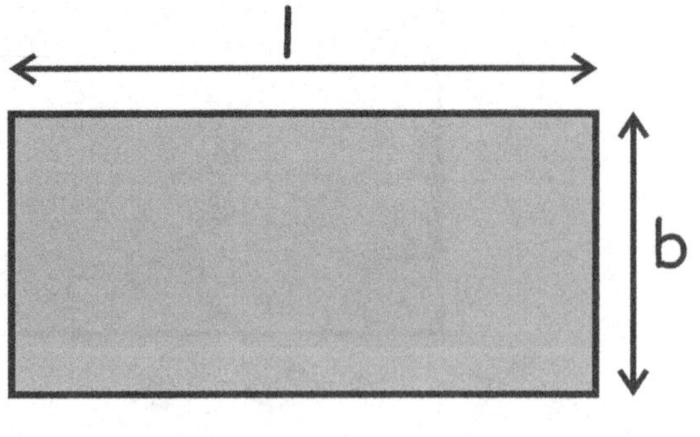

Area = l b x

Perimeter = 2(l + b)

8. Acute angle

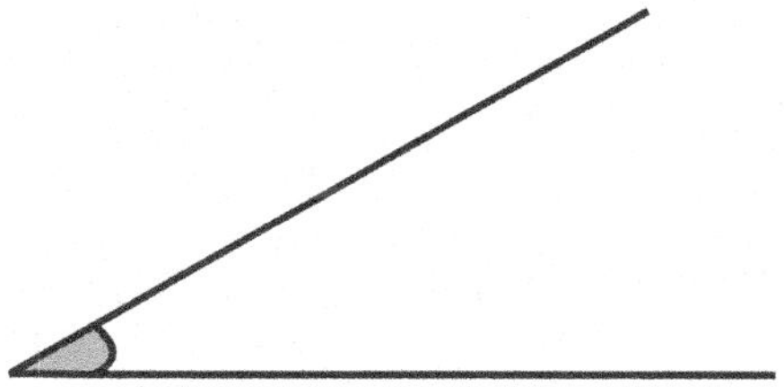

Any angle less than 90°
is called as acute angled

9. Right angle

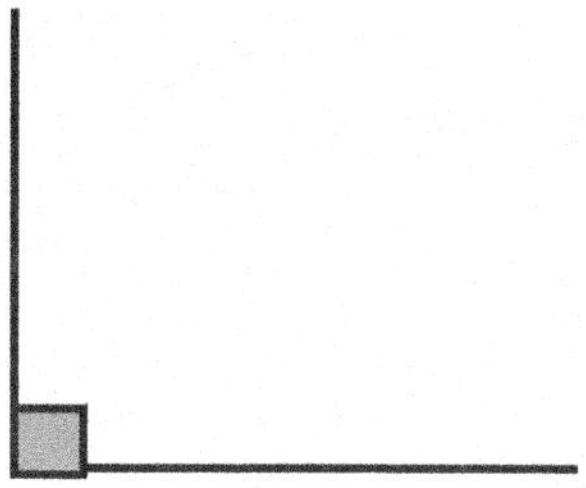

An angle equal to 90°
is called as right angled

10. Obtuse angle

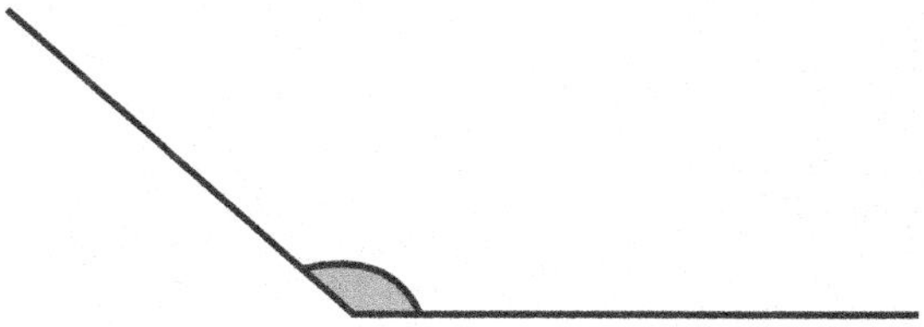

Any angle greater than 90°
is called as obtuse angled

11. Perimeter of a polygon KLMNOP

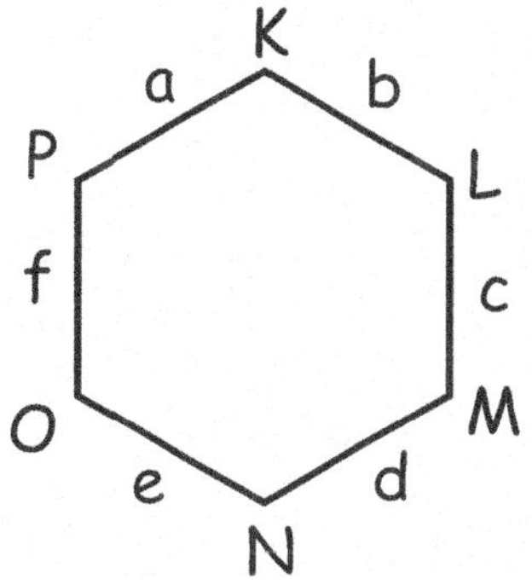

Perimeter =
(a + b + c + d + e + f) units

Abbreviations

milligram	mg
gram	g
kilogram	kg
milliliter	mL
liter	L
kiloliter	kL
millimeter	mm
centimeter	cm
meter	m
kilometer	km
square centimeter	cm^2
cubic centimeter	cm^3

year	yr
month	mon
hour	hr
minute	min
second	sec

volume	V
total Square Area	S.A
area of base	B
ounce	oz
pound	lb
quart	qt
gallon	gal.
inches	in.
foot	ft
yard	yd
mile	mi.
square inch	sq in.
square foot	sq ft
cubic inch	cu in.
cubic foot	cu ft

Grade 3
SOL
Practice Test - 4

1. Rik sold 209,493 tickets for his high school game Carnival in Spring. What is 209,493 rounded to the nearest hundred thousand ?

   (A) 209,000

   (B) 209,490

   (C) 209,400

   (D) 200,000

2. Which of the below shows "Seventy seven thousand, nine hundred seven" in standard form ?

   (A) 77,000,907

   (B) 777,097

   (C) 77,907

   (D) 777,907

3. Which of the below statement is true ?

   (A) $66666 < 666666$

   (B) $66666 > 66667$

   (C) $66666 = 6666$

   (D) $66667 > 666666$

4. Which of the below statement is true ?

(A) 92,960 = 99,960 (B) 92,960 > 93,990

(C) 92,960 > 92,990 (D) 92,960 > 91,960

5. What is 101,115 rounded to the nearest hundred ?

(A) 101,115 (B) 100,110

(C) 101,100 (D) 100,000

6. The adjacent figure represents one whole part

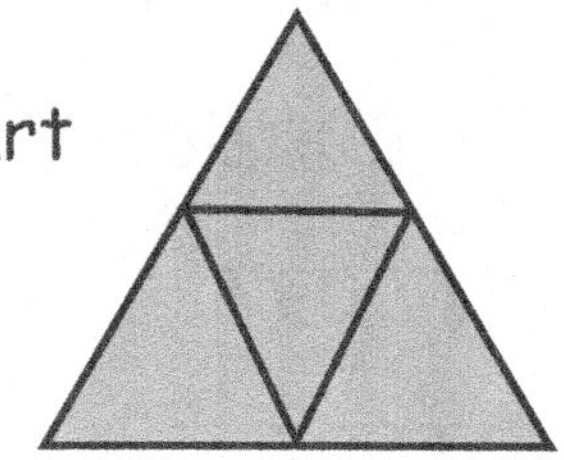

Then what number is represented by the below ?

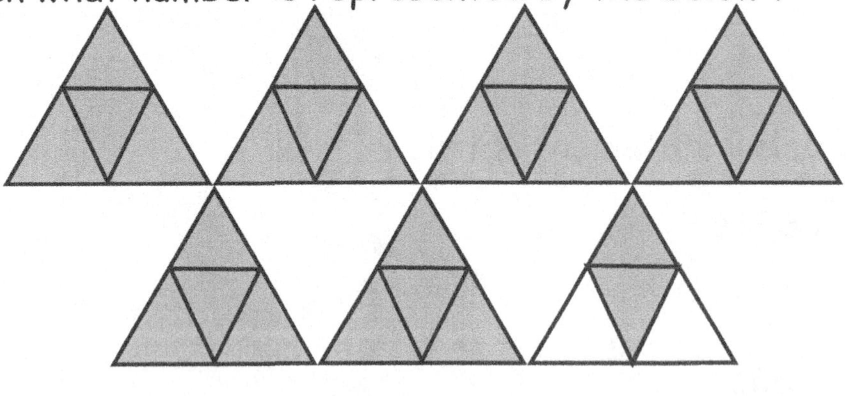

(A) $1\frac{2}{3}$ (B) $5\frac{3}{4}$

(C) $6\frac{1}{2}$ (D) $7\frac{4}{4}$

7. The figure below is shaded to represent the number 1.

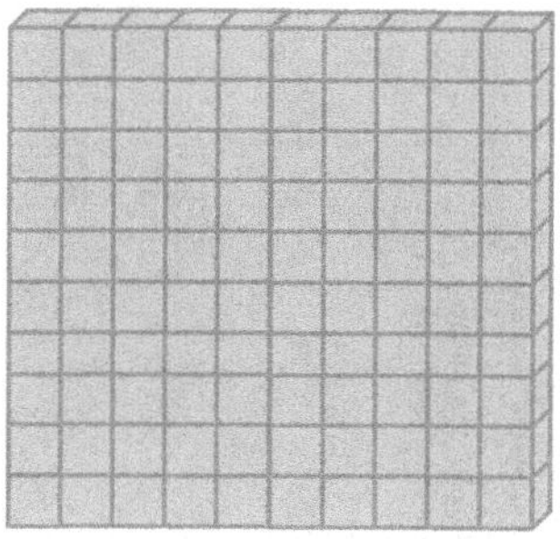

The below model represents which number ?

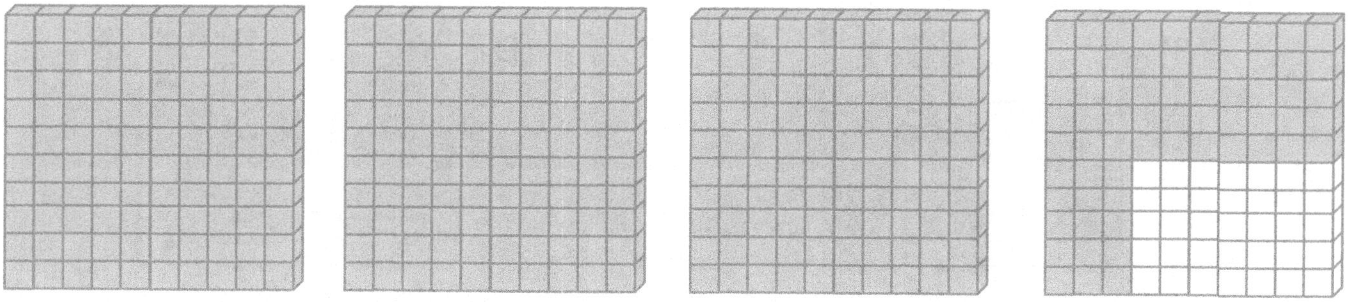

(A) 3.55 (B) 3.70 (C) 3.64 (D) 3.65

8. This model is shaded to show a fraction of a whole.

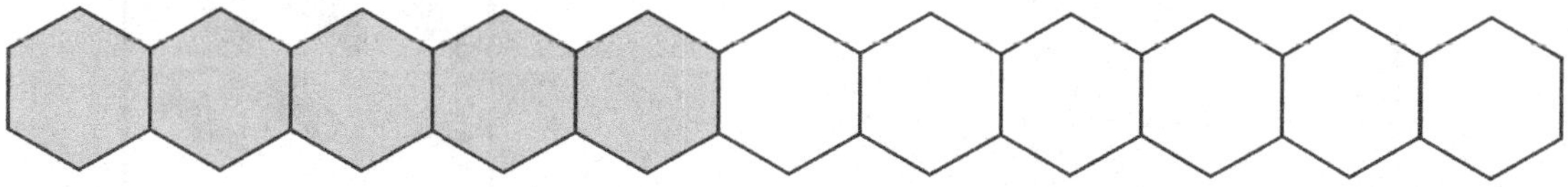

Which of the following is shaded to show a fraction with a value GREATER THAN the one above ?

(A) 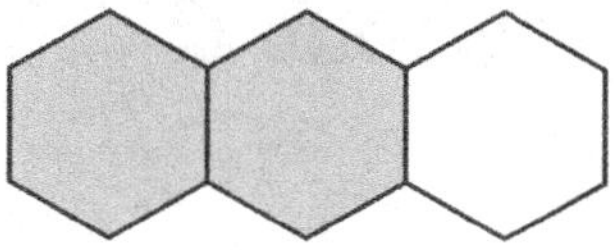

(B) 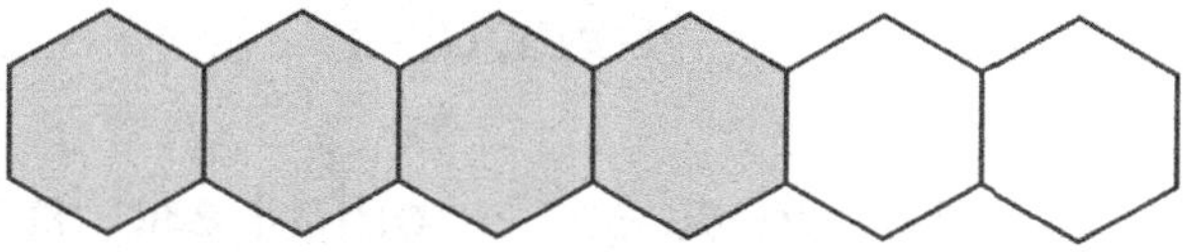

(C) 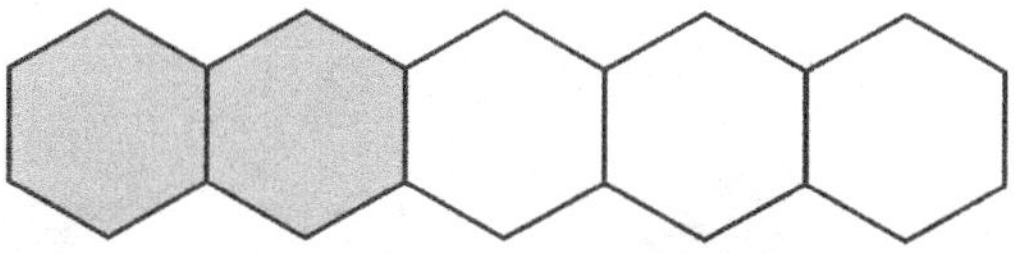

(D) 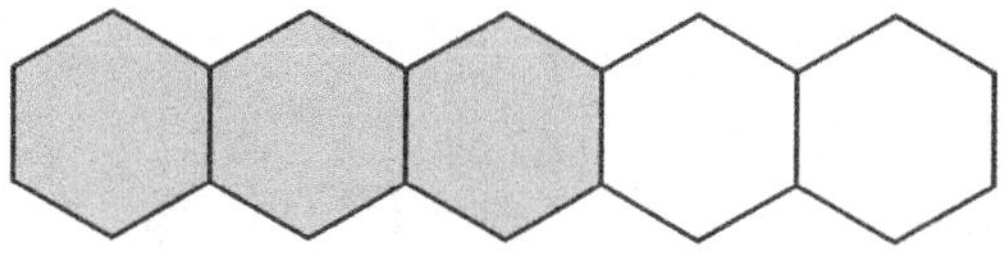

9. Which picture box below shows equal number of  and  ?

(A) 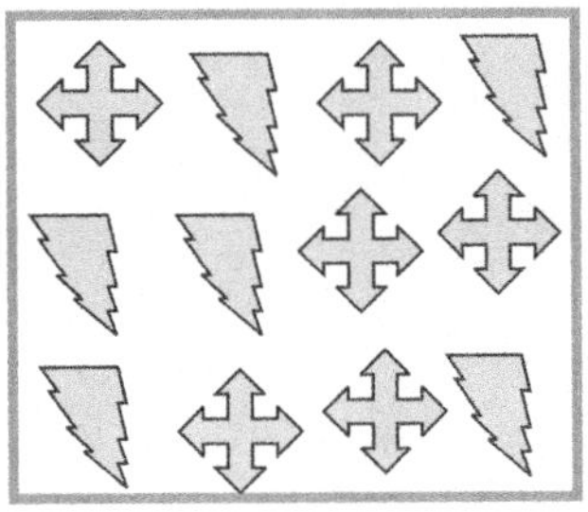

(B) 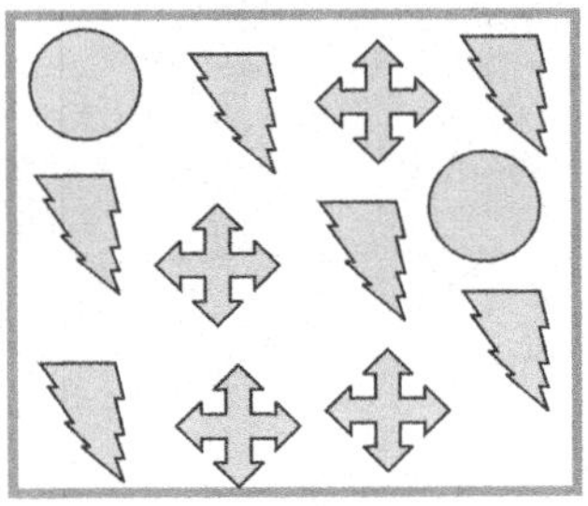

(C) 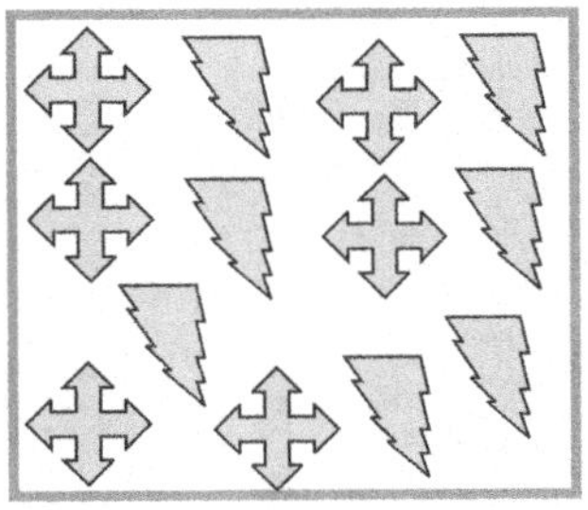

(D) 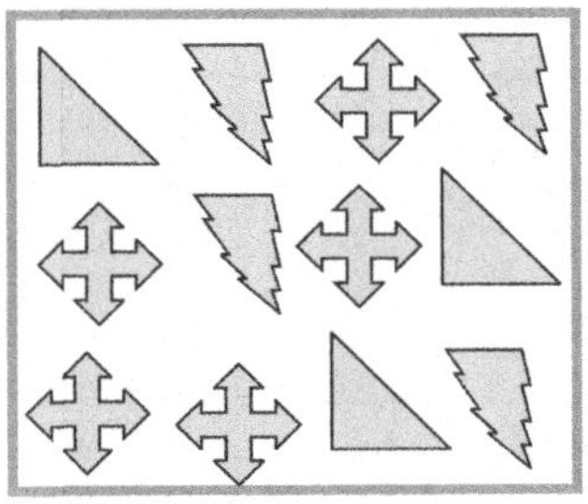

10. Lena has two types of sweets for her snack. She sorted them into the following two groups. She placed sweets of the same type and flavour into a group.

Group L

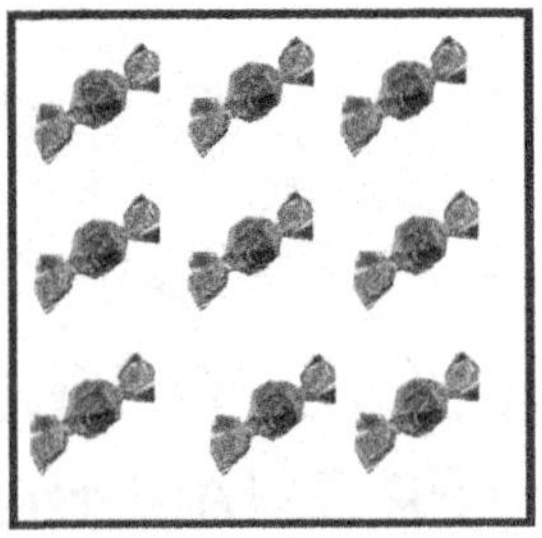

Group C

Which sweet should Lena include in Group C ?

(A)

(B) 

(C) 

(D) 

11. Cathy and her team made 9,758 Necklaces to sell for a school fundraiser. What is the value of 7 in 9,758 ?

(A) 70

(B) 7,000

(C) 70,000

(D) 700

12. This  represents a value of 1.

Which model represents the value of 9 in 64,598 ?

(A) 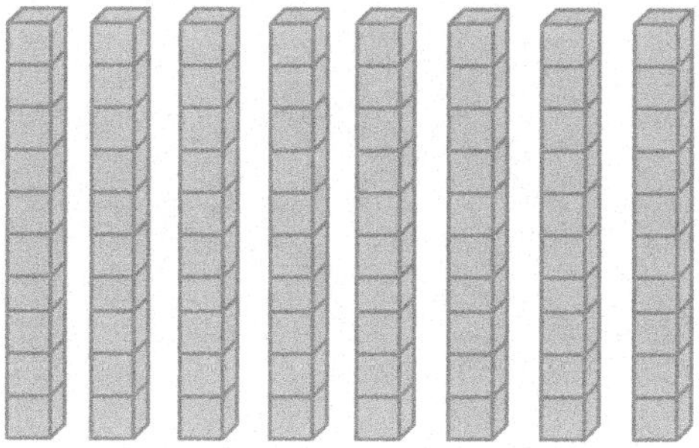

(B) 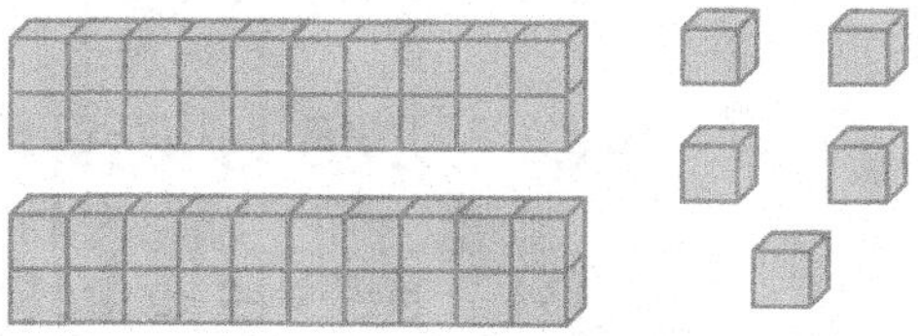

(C) 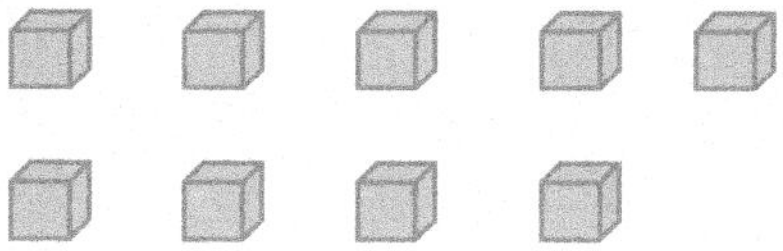

(D) 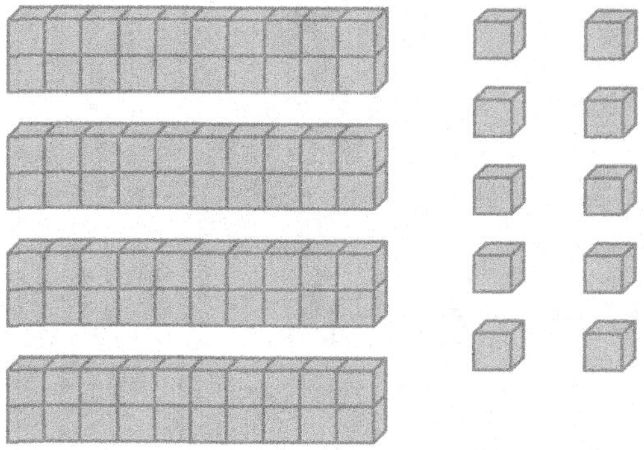

13. What fraction of Oranges are there from the below group of fruits ?

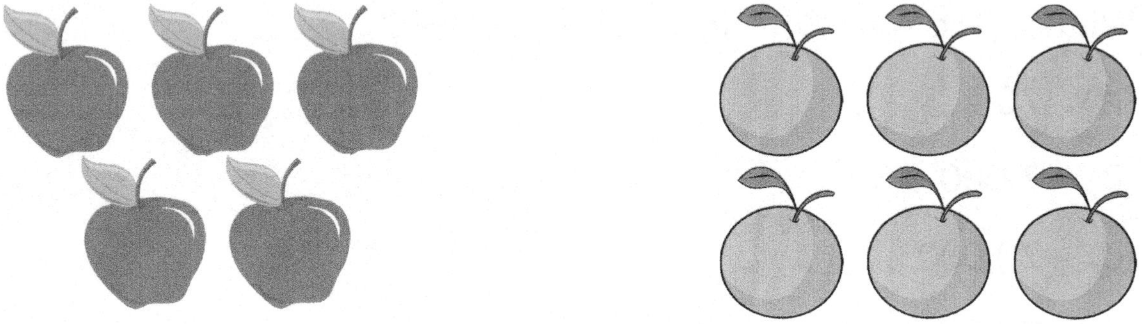

(A) $\frac{6}{11}$ (B) $\frac{5}{12}$ (C) $\frac{6}{12}$ (D) $\frac{5}{11}$

14. Which of the below statement is true ?

(A) 504 is greater than 5003

(B) 912 is less than 670

(C) 275 is less than 999

(D) 838 is equal than 1838

15. Amy has 4 children. She gives 6 cookies to each one of them. How many cookies did she gave them in all ?

(A) 24

(B) 6

(C) 10

(D) 14

16. Evaluate $44 \div 2 = \boxed{?}$

(A) 12

(B) 28

(C) 22

(D) 9

17. Gary has 16 pencils and 9 crayons.

16	9

How many more pencils does Gary has than crayons ?

(A) 11

(B) 7

(C) 10

(D) 12

18. Evaluate $2003 + 199 = \boxed{?}$

(A) 2,202

(B) 2,210

(C) 2,209

(D) 2,212

19. This model is shaded to show 1 whole.

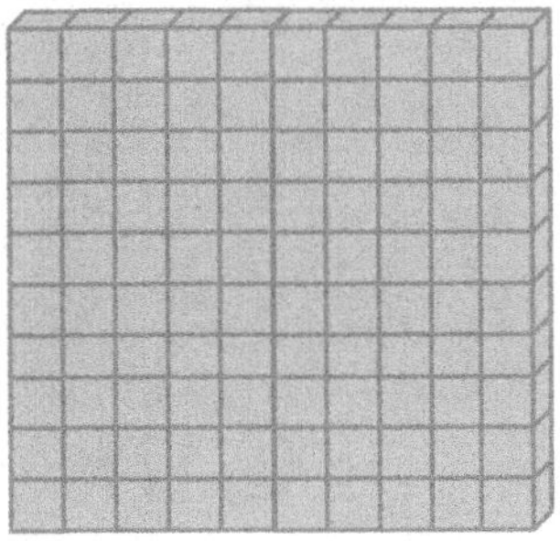

What decimal is shown by the model below ?

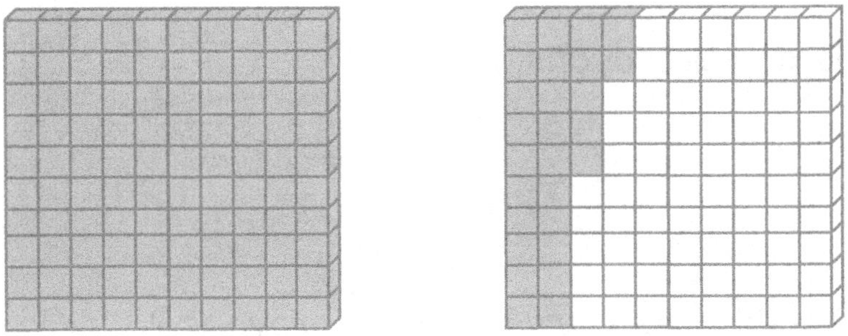

(A) 1.17 (B) 1.22 (C) 1.67 (D) 1.27

20. Evaluate

$$\begin{array}{r} 7.26 \\ +2.23 \\ \hline \end{array}$$

____

(A) 10.9

(B) 9.49

(C) 12.3

(D) 25.8

21. This model is shaded to show one whole.

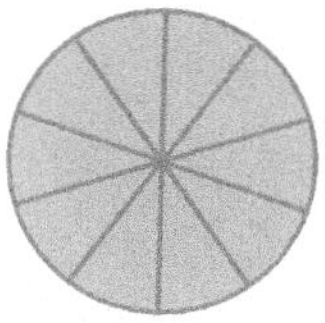

The two models below are each shaded to show a fraction

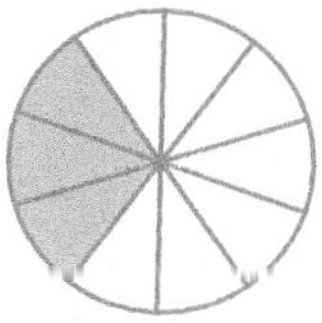

Model 1

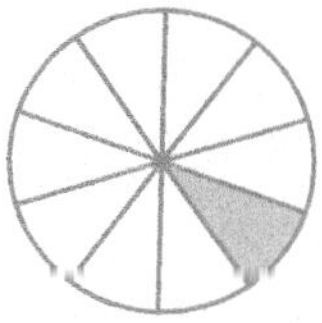

Model 2

One fraction is subtracted from the other. Which model is shaded to show the difference ?

(A) $\frac{1}{5}$ (B) $\frac{1}{10}$ (C) $\frac{1}{15}$ (D) $\frac{1}{20}$

22. Evaluate

6 x 6 = ?

(A) 12

(B) 10

(C) 36

(D) 90

23. Evaluate $7 \times 2 = \boxed{?}$

(A) 12

(B) 10

(C) 9

(D) 14

24. Evaluate $5.9 + 2.5 = \boxed{?}$

(A) 7.7

(B) 8.4

(C) 6.9

(D) 7.07

25. Which of the figures below appears to have a line of symmetry ?

(A) 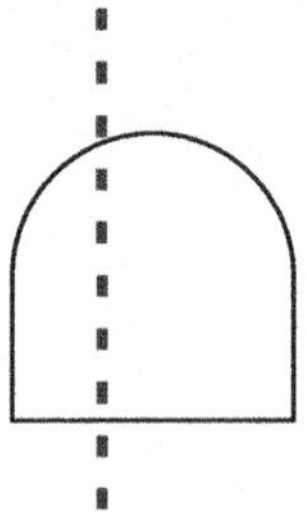

(B) 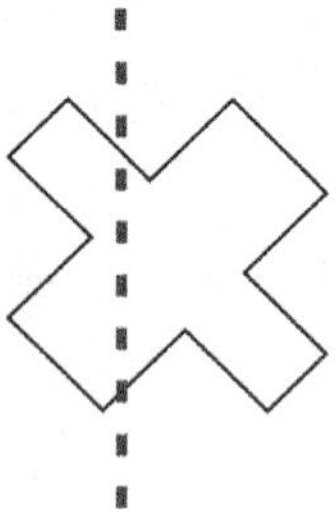

(C) 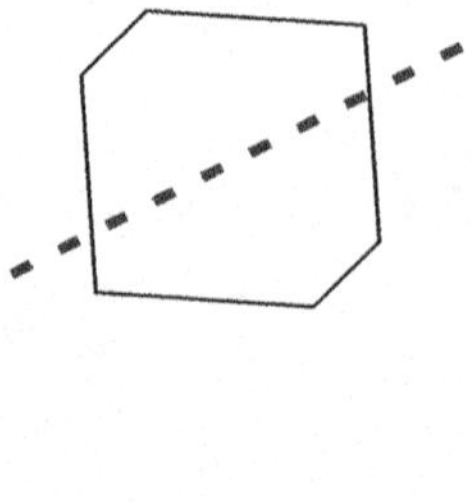

(D) 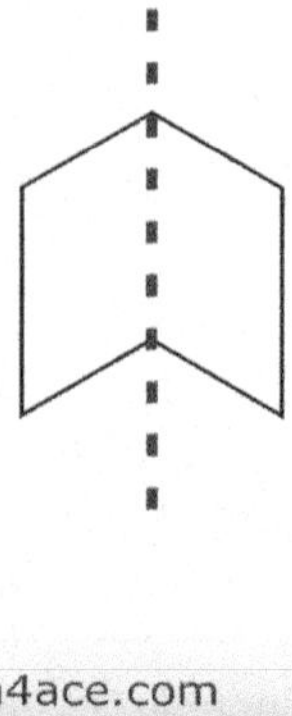

26. Dan makes 52 glasses of Lemonade. He sells 41 glasses of the Lemonade. What fractional part of the Lemonade was left ?

(A) $\frac{10}{52}$ (B) $\frac{11}{52}$ (C) $\frac{8}{52}$ (D) $\frac{15}{52}$

27. Water in the fish tank can be measured in which units ?

(A) Miles

(B) Grams

(C) Gallons

(D) Pounds

28. Jack has 8 more weeks of school before summer vacation. How many days are left before the summer vacation ?

(A) 56

(B) 30

(C) 16

(D) 64

29. Identify the closest temperature shown in the below figure (Approximately)

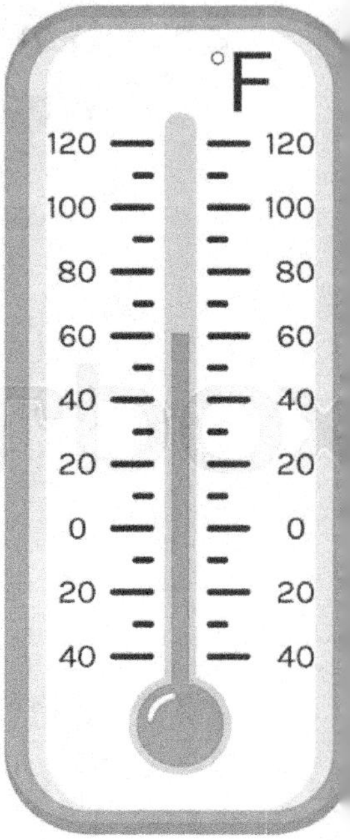

(A) 49^{0} Fahrenheit

(B) 150^{0} Fahrenheit

(C) 26^{0} Fahrenheit

(D) 60^{0} Fahrenheit

30. Identify the picture below?

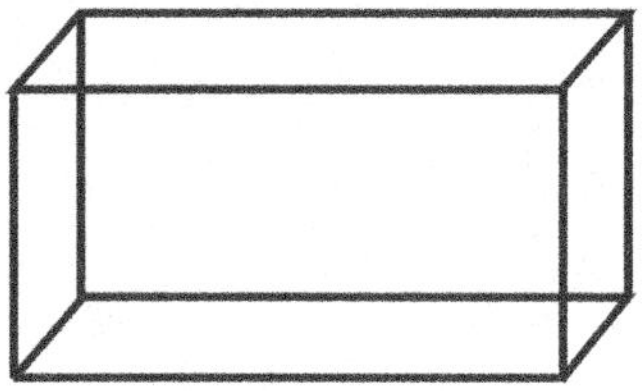

(A) Triangle

(B) Prism

(C) Cuboid

(D) Circle

31. Length of one tooth pick is equal to 1 cm. What is the closest length of the soccer ball?

(A) 3 cm

(B) 5 cm

(C) 10 cm

(D) 8 cm

32. Kate is wondering which of the below is showing the time of 3.20 PM ?

(A) 

(B) 

(C) 

(D) 

33. Identify which of the following figures has the line of symmetry across the dotted line ?

(A) 

(B) 

(C) 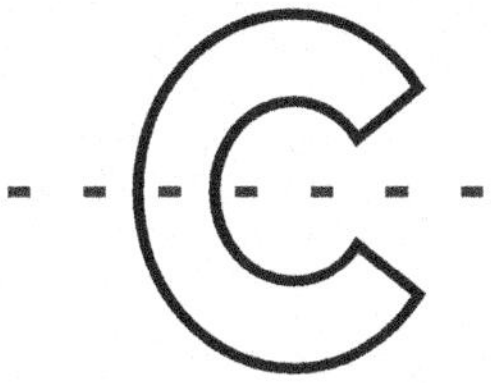

(D) 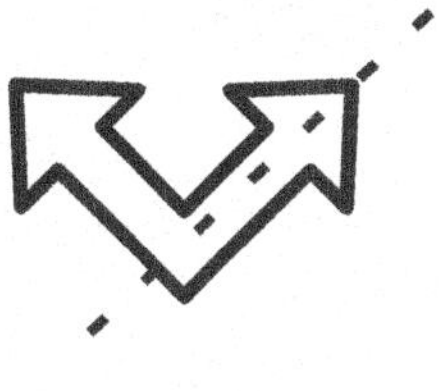

34. A Rhombus can also be a.....?

(A) Parallelogram

(B) Trapezoid

(C) Circle

(D) Triangle

35. Find the perimeter of the hexagon with equal side length as given below ? (Approximately using the ruler)

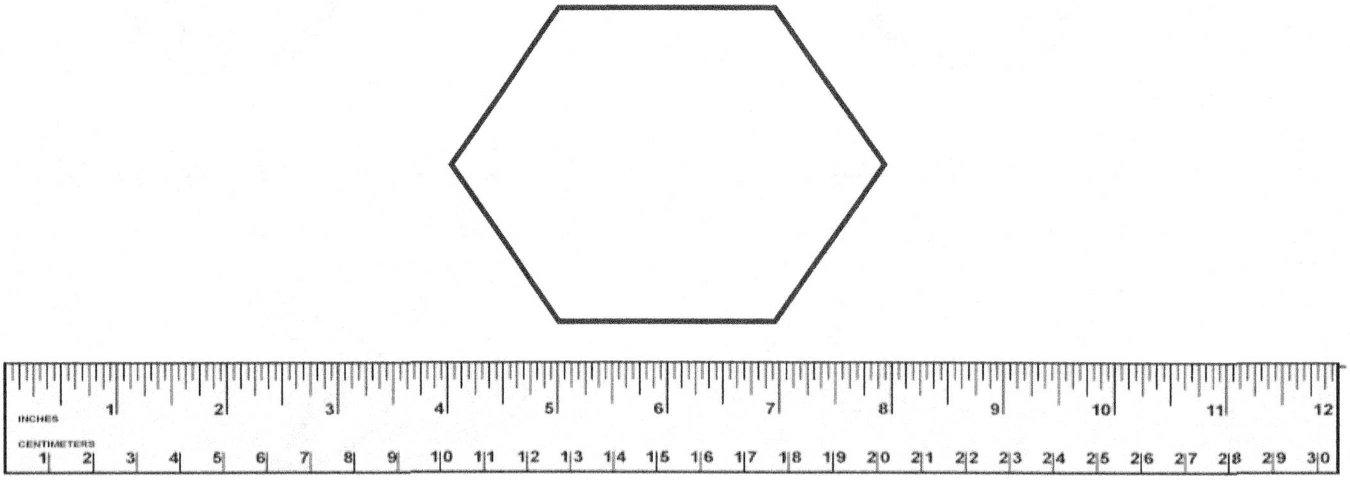

(A) 12.5 inches

(B) 16 inches

(C) 12 inches

(D) 14.5 inches

36. Tom has

He saved this week

How much money did he have in all ?

(A) $ 4.75

(B) $ 5.66

(C) $ 6.55

(D) $ 4.25

37. Below data shows the number of houses sold in Twin Town county this quarter ?

Month	January - 2019	February - 2019	March - 2019
Number Sold	48	32	56

Cathy prepares the below chart representing the above data

Month	Number Sold
January - 2019	
February - 2019	
March - 2019	

The symbol  represents the value of ____________ ?

(A) 3

(B) 4

(C) 2

(D) 8

38. Beth has a collection of animal stickers as shown below

Which bar graph correctly shows the number of each kind of stickers ?

(A)

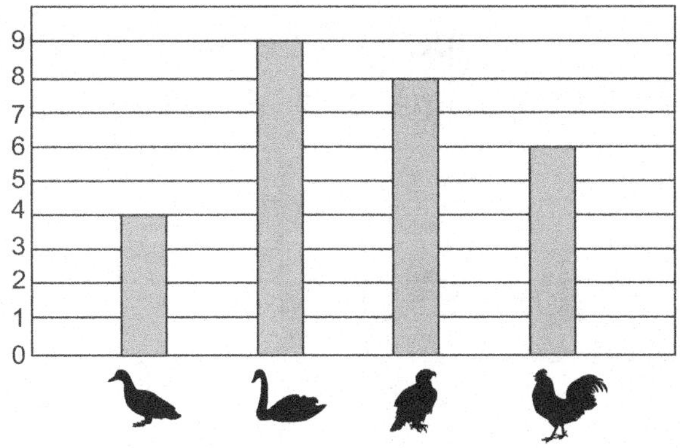

(B)

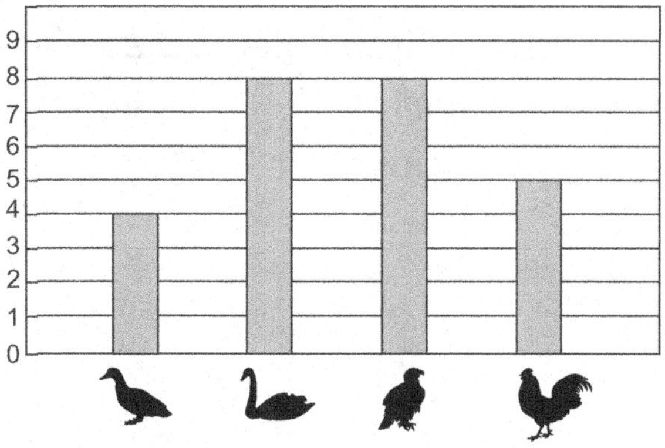

(C)

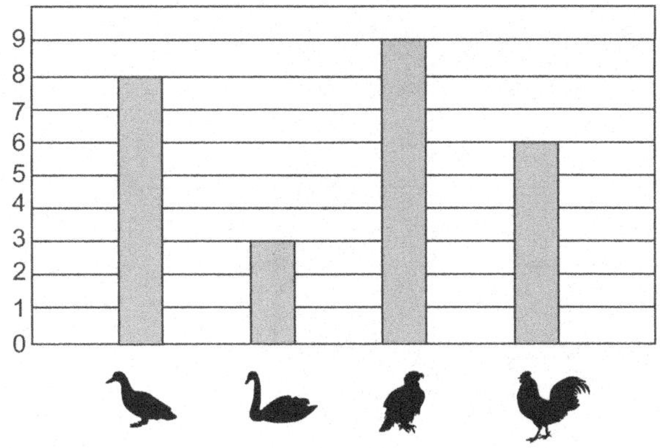

(D)

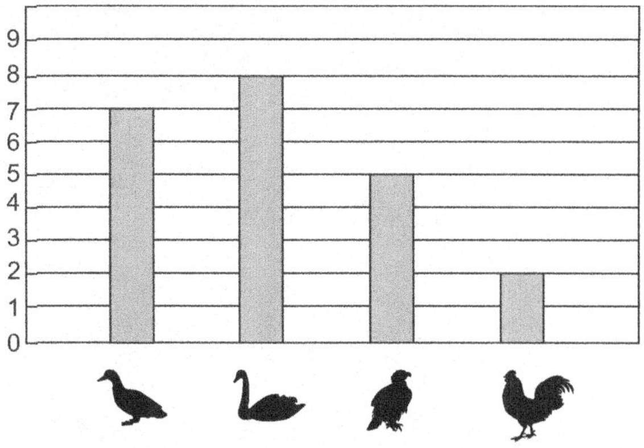

39. Below numbers are written on individual cards and are placed in a jar after carefully placing each of them in envelops. Rik picks a card from the jar. What is the chance of the number in the envelop being a prime number ?

(A) Impossible

(B) Likely, but not certain

(C) Unlikely, but not impossible

(D) Certain

40. Alice coloured small squares in the shape of an alphabet from the below grid. If □ is equal to one square centimeters. What is the area of the alphabet "S" ?

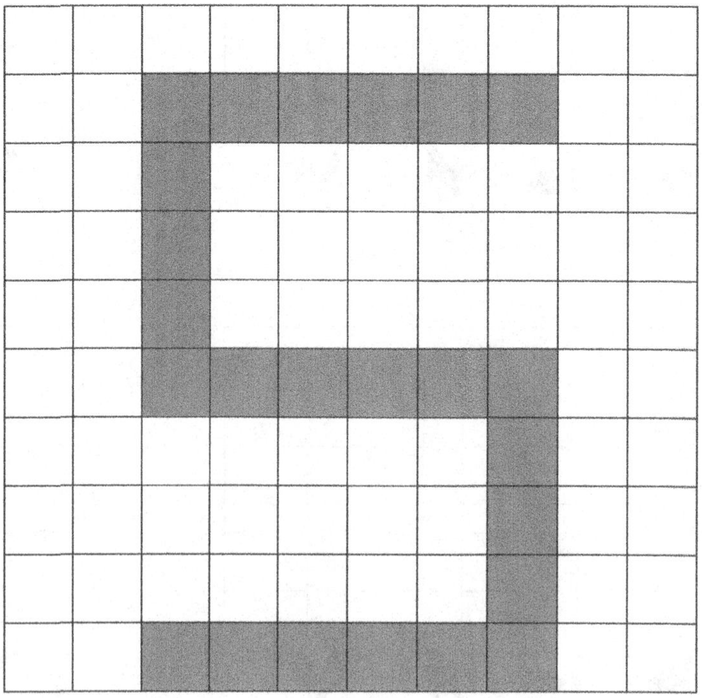

(A) 20 (B) 19 (C) 24 (D) 15

41. Esha uses the math fact $4 \times 3 = 12$ to solve a math problem. Which of the below could be the problem she is trying to solve.

(A) $\square \times 4 = 3$

(B) $\square \div 3 = 4$

(C) $\square + \square = 7$

(D) $12 - \square = 4$

42. Find the next number in the series ?

16 , 20 , 24 , 28 , 32,.....?

(A) 36 (B) 33 (C) 40 (D) 39

43. Which number can be placed on the line to make the number sentence true ?

987 X ? = 0

(A) 1 (B) 10 (C) 100 (D) 0

44. Using the skip counting rule what is the next number in the series ?

515 , 520 , 525 , 530 , 535 , 540 , 545 ,... ?

(A) 600 (B) 575 (C) 550 (D) 500

45. The table shows the total costs of different Art pieces that Dora is selling in an art exhibition.

Pencil Costs

Number of Art Pieces	Total Cost
1	$ 0.07
2	$ 0.14
3	$ 0.21
4	$ 0.28
5	$ 0.35
6	$?

Based on the pattern in the table, what will be the total cost for 6 books ?

(A) $ 0.42 (B) $ 0.36 (C) $ 0.40 (D) $ 0.50

46. Grade 3 teachers from H wood school took a survey in their respective classes on their favourite sport soccer. Below graph represents the number of votes received.

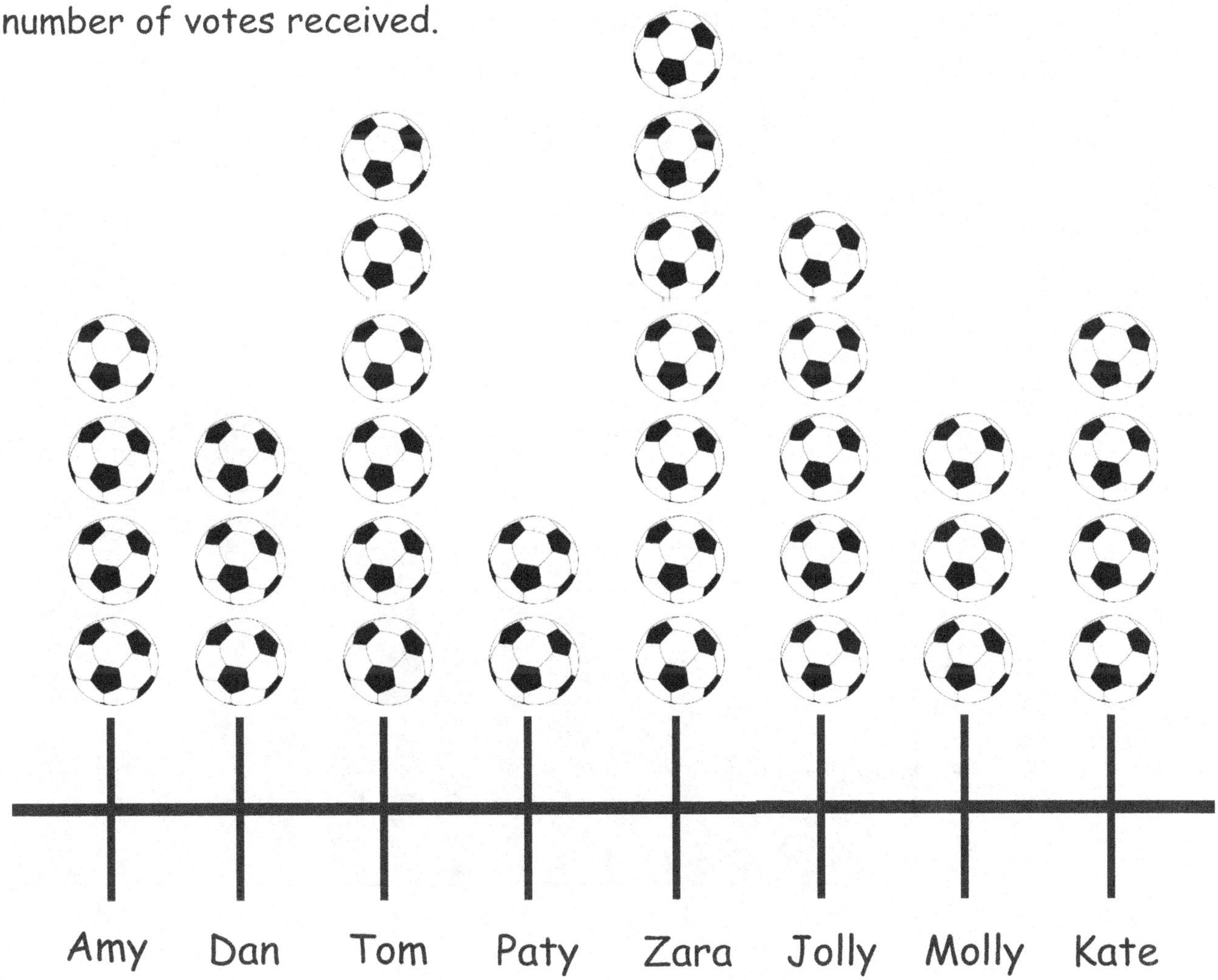

Third Grade Teachers

Which class has exactly six students who likes to play soccer ?

(A) Kate (B) Zara

(C) Jolly (D) Tom

47. Grade 3 of Trendy Blazers elementary school collects story books for fifteen weeks for a book donation project. How many books were collected by Trendy Blazers elementary school ?

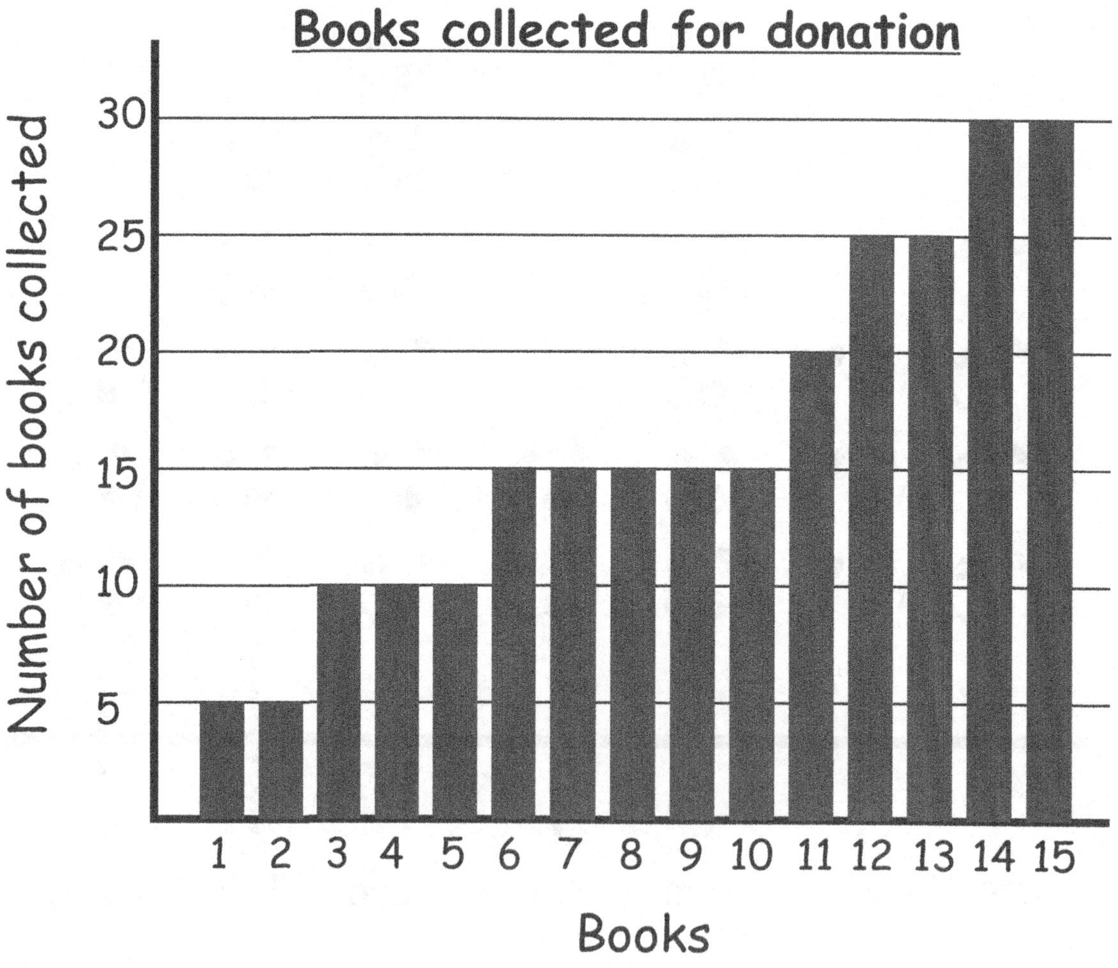

(A) 385 (B) 275

(C) 500 (D) 245

48. Super Sonic stem club participated in various competitions for last ten years. Below graph represents the number of awards the club won across the years. What is missing on the graph ?

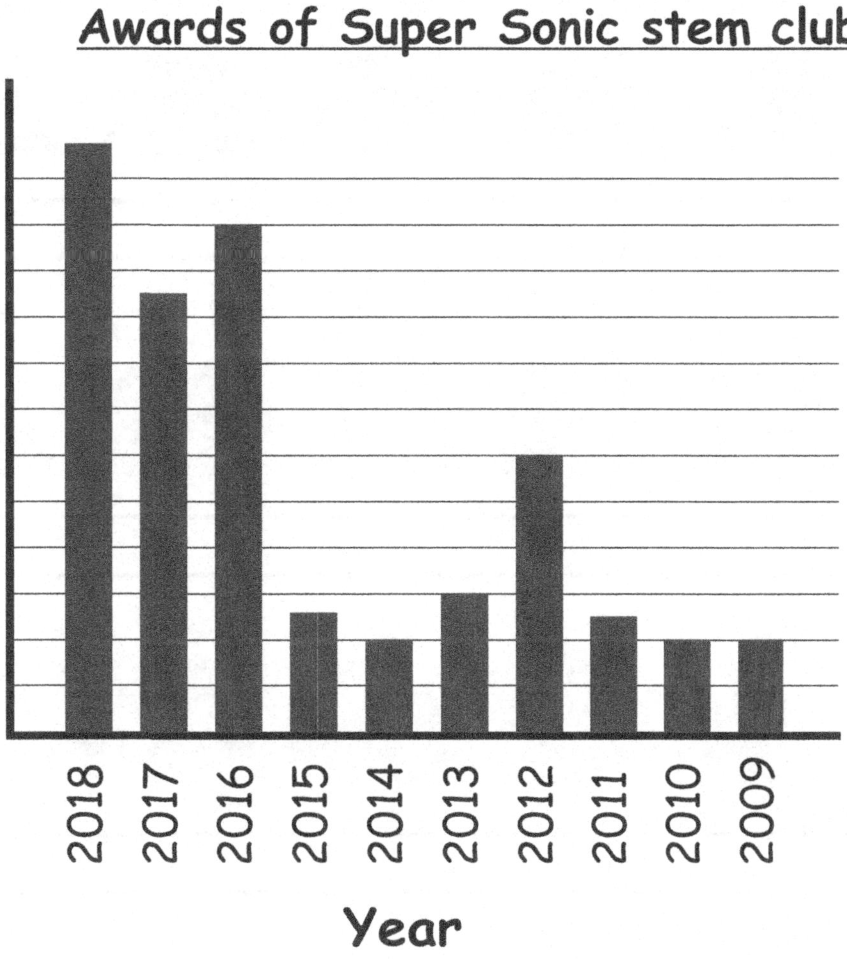

(A) First place award

(B) Third place award

(C) Number of awards

(D) Second place award

49. Kyla sorts her dresses as a group of six. She has 24 dresses all. Which of her below grouping shows as a multiple fact ?

(A)

(B)

(C)

(D) 

50. Identify the picture below ?

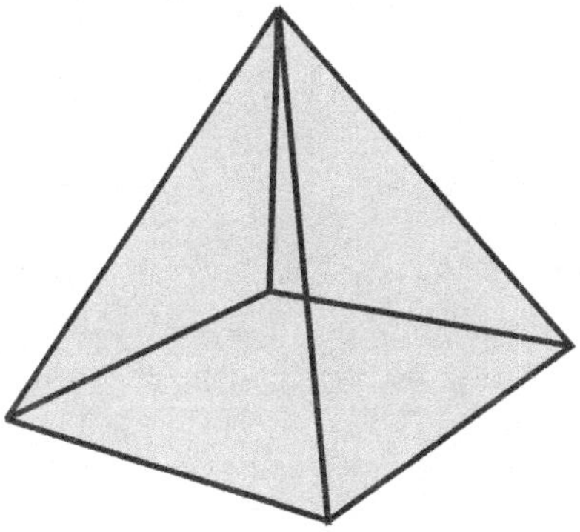

(A) Prism
(B) Cone
(C) Pyramid
(D) Triangle

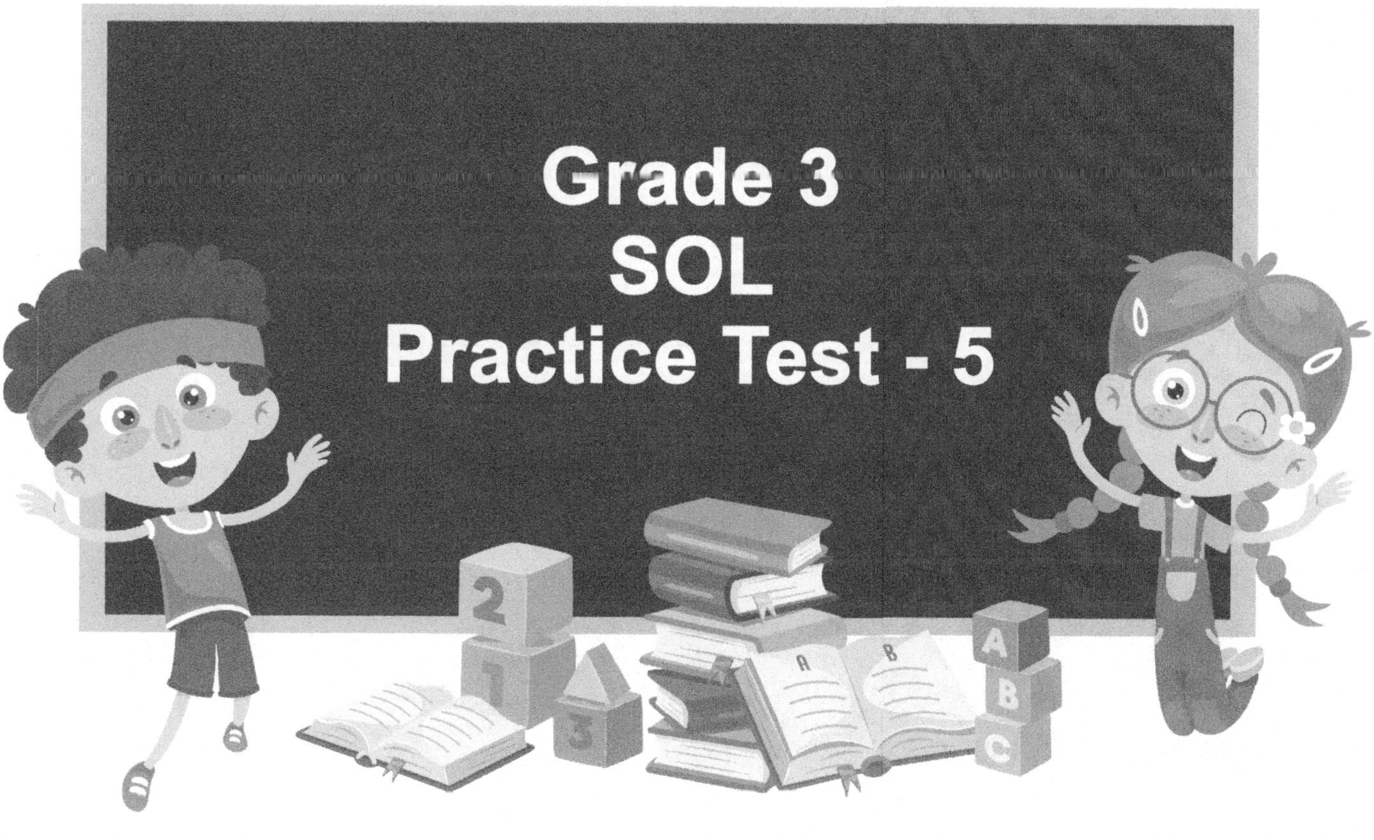
Grade 3
SOL
Practice Test - 5

1. Lily sold 1,509,099 bracelets this year at her store. What is 1,509,099 rounded to the nearest hundred thousand ?

    (A) 2,509,000

    (B) 2,000,000

    (C) 2,500,000

    (D) 1,500,000

2. Which of the below shows " Eight hundred six thousand, four hundred five" in standard form ?

    (A) 8,006,415

    (B) 816,415

    (C) 806,405

    (D) 806,450

3. **24 - 5 = 19**

    The number sentence given above can also be written as ?

    (A) 19 + 5 = ____

    (B) 19 + 24 = ____

    (C) 19 - 5 = ____

    (D) 24 + 5 = ____

4. Which of the below statement is true ?

(A) $9{,}409 > 9{,}574$

(B) $9{,}409 < 9{,}409$

(C) $9{,}409 > 9{,}037$

(D) $9{,}409 > 9{,}734$

5. If the below figure represents one whole part

then what frational number is represented by the below ?

(A) $5\frac{2}{5}$

(B) $4\frac{3}{8}$

(C) $6\frac{2}{4}$

(D) $7\frac{4}{4}$

6. The figure below is shaded to represent the number 1.

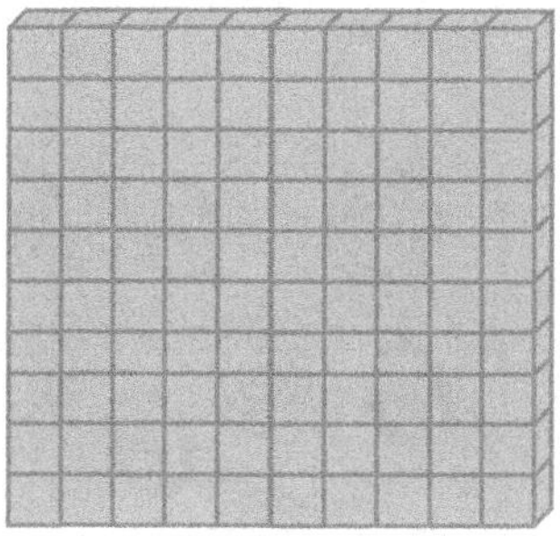

The below model represents which number ?

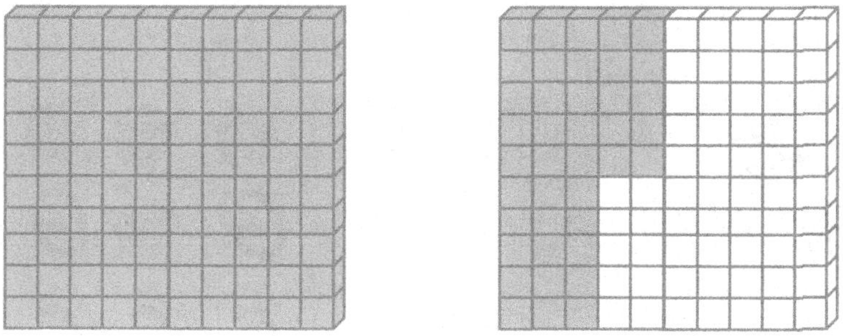

(A) 1.34 (B) 2.80 (C) 2.08 (D) 1.40

7. This model is shaded to show a fraction of a whole.

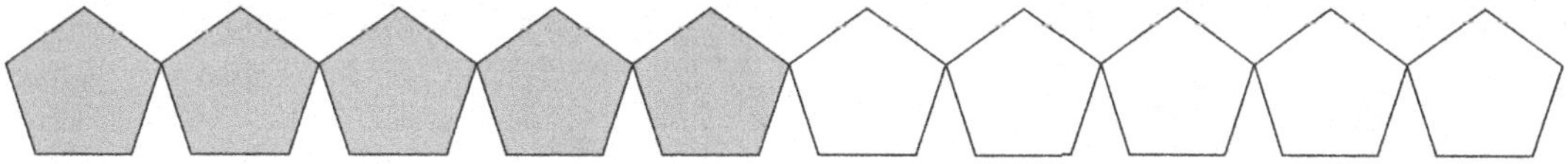

Which of the following is shaded to show a fraction with a value GREATER THAN the one above ?

(A) 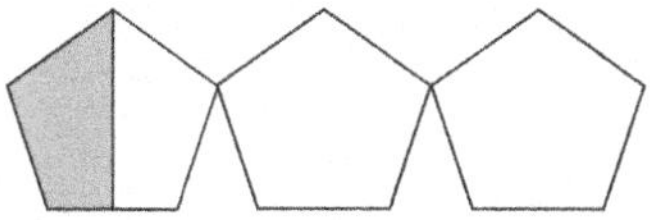

(B) 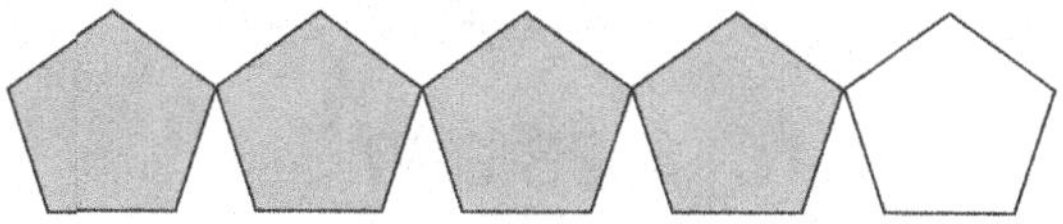

(C) 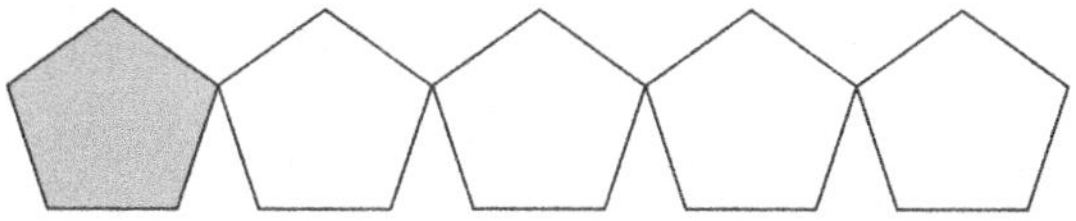

(D) 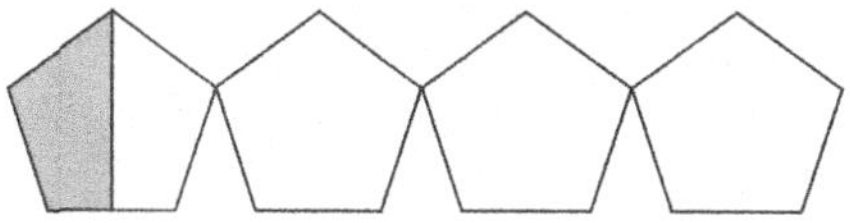

8. Which picture box below has more letters of "G"?

(A)

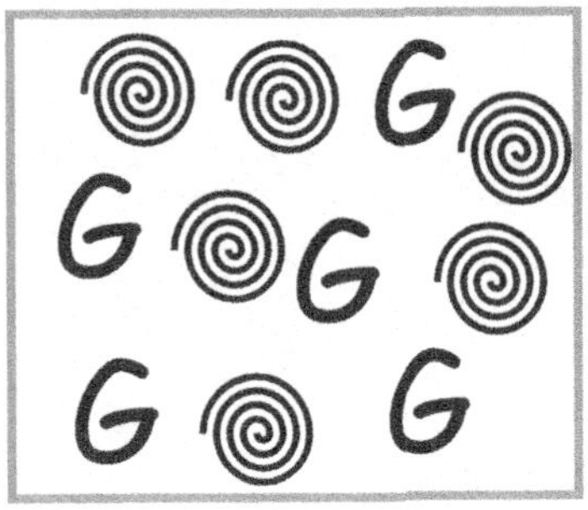

(B)

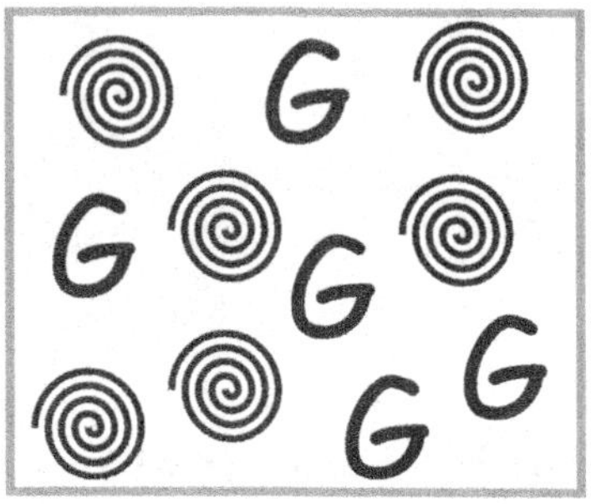

(C)

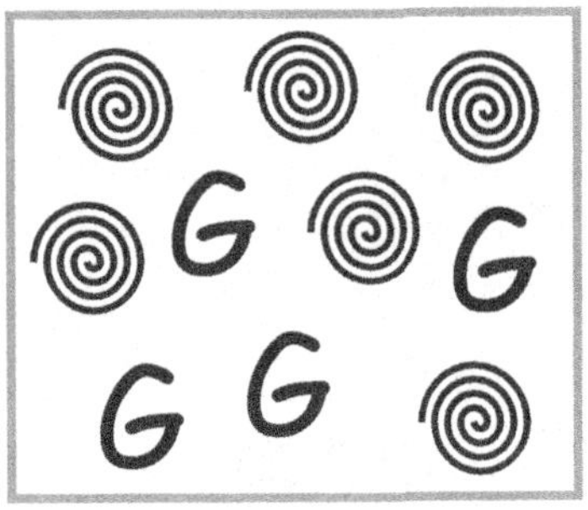

(D)

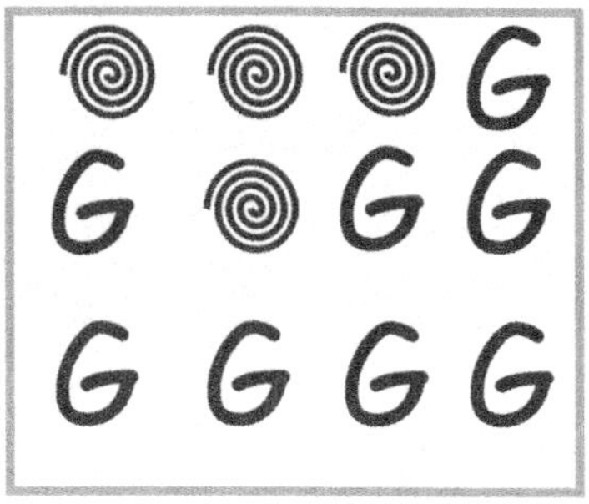

9. What is 23,970 rounded to the nearest ten thousand ?

(A) 20,000

(B) 23,900

(C) 23,000

(D) 24,000

10. Zach solved 5,031 games in math problems this year. What is the value of three in 5,031 ?

(A) 30,000

(B) 3,000

(C) 30

(D) 300

11. This  represents a value of 1.

Which model represents the value of 5 in 861, 256 ?

(A) 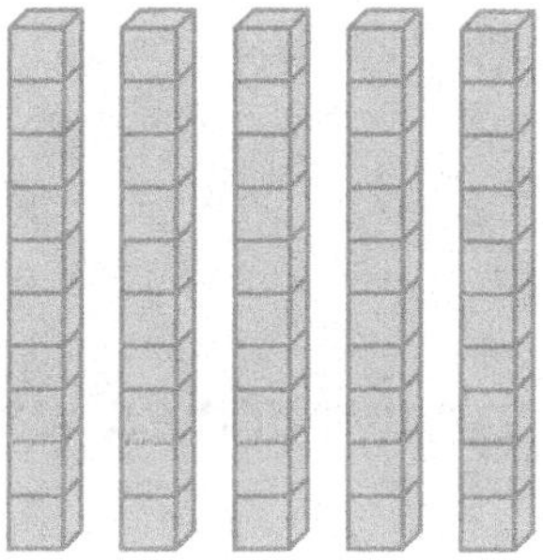

(B) 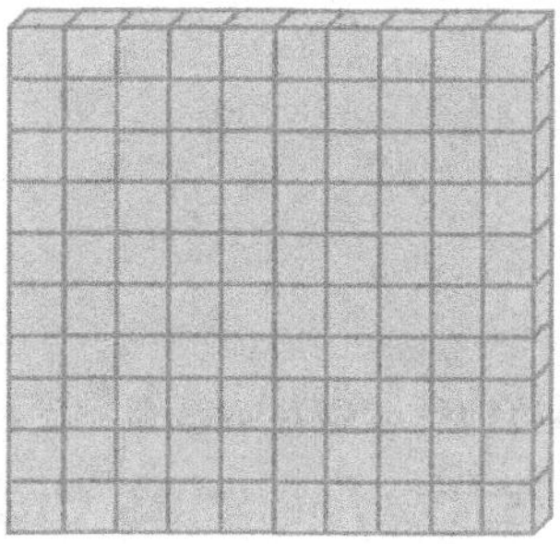

(C) 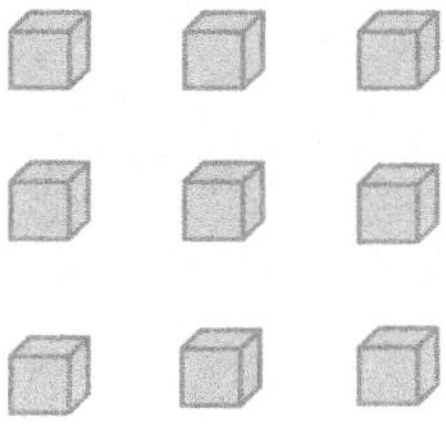

(D) 

12. What fraction of helmets are there from the below ?

(A) $\frac{5}{8}$ (B) $\frac{5}{13}$ (C) $\frac{3}{5}$ (D) $\frac{8}{13}$

13. Evaluate $80 \div 10 = \boxed{?}$

(A) 10

(B) 8

(C) 9

(D) 800

14. Jack gets two pounds of strawberries. Five strawberries weigh a pound. How many strawberries did he got in all ?

(A) 14

(B) 5

(C) 7

(D) 10

15. Evaluate $18 \times 1 = \boxed{?}$

(A) 71

(B) 60

(C) 18

(D) 19

16. Jason has 42 X - box games and 71 Play station games

42	71
X - box games	Play station games

How many more Play Station games does Jason have than X - box games ?

(A) 29

(B) 121

(C) 1023

(D) 21

17. Evaluate $9,438 + 491 = \boxed{?}$

(A) 8,090 (B) 8,100 (C) 8,108 (D) 9,929

18. This model is shaded to show 1 whole.

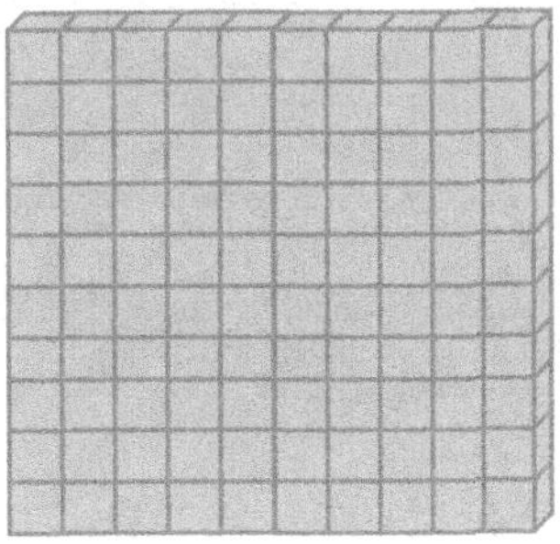

What decimal is shown by the model below ?

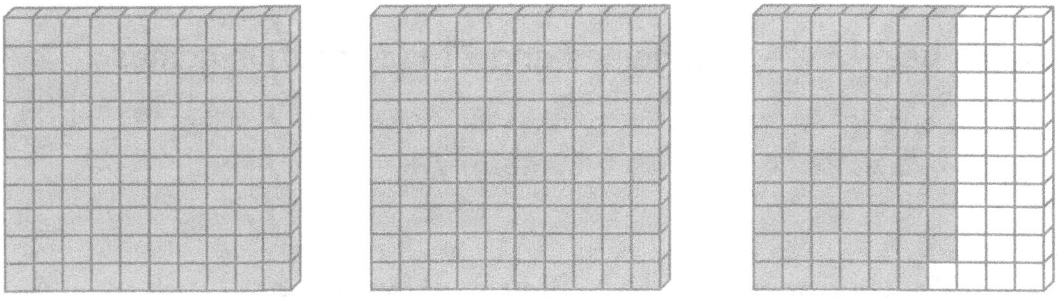

(A) 2.59 (B) 2.70 (C) 2.69 (D) 2.49

19. Evaluate

$$\begin{array}{r} 5.57 \\ -\ 3.21 \\ \hline \end{array}$$

(A) 2.36

(B) 8.78

(C) 1.76

(D) 2.77

20. This model is shaded to show one whole.

The two models below are each shaded to show a fraction

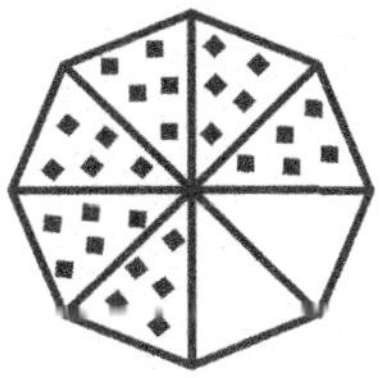

Model 1

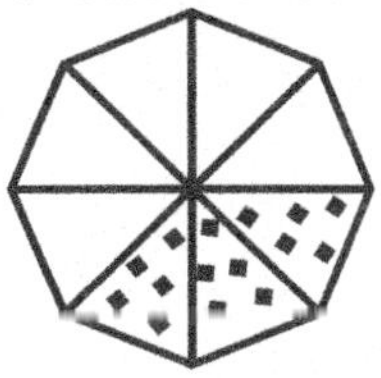

Model 2

One fraction is subtracted from the other. Which model is shaded to show the difference ?

(A) $\frac{1}{2}$ (B) $\frac{13}{8}$ (C) $\frac{1}{3}$ (D) $\frac{3}{8}$

21. Evaluate

$6 \times 6 = \boxed{?}$

(A) 36

(B) 10

(C) 01

(D) 90

22. The below number sentence shows a multiplication property.

$4 \times 6 = 6 \times 4$

Which of the following number sentence shows the same property as above ?

(A) $12 \times 3 = 20 \times 9$

(B) $10 = 5 \times 2$

(C) $9 \times 3 = 6 \times 9$

(D) $18 \times 1 = 1 \times 18$

23. Evaluate $15.7 + 1.2 = \boxed{?}$

(A) 17.7 (B) 17.8 (C) 16.9 (D) 17.07

24. Which of the figures below appears to have a line of symmetry ?

(A) 

(B) 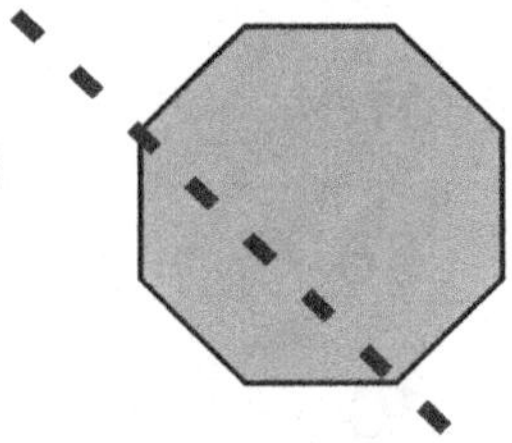

(C) 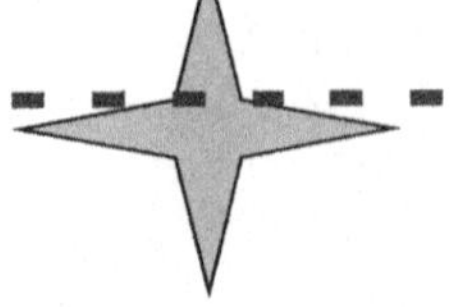

(D) 

25. Danny divided a large pizza into 8 equal pieces. He gave 2 pieces to his sister and ate 3 pieces himself. What fractional part of the pizza was left ?

(A) 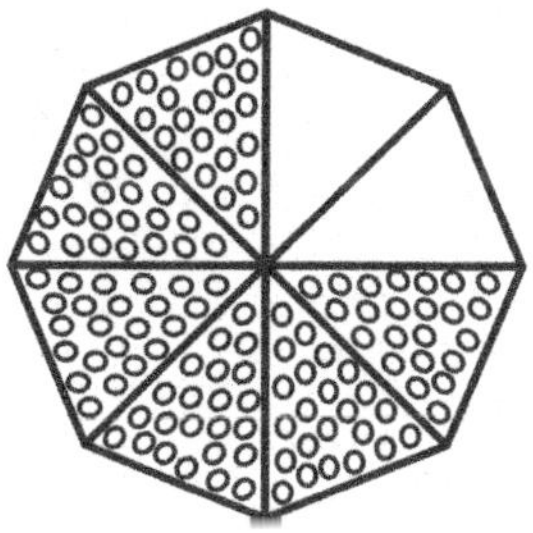

(B) 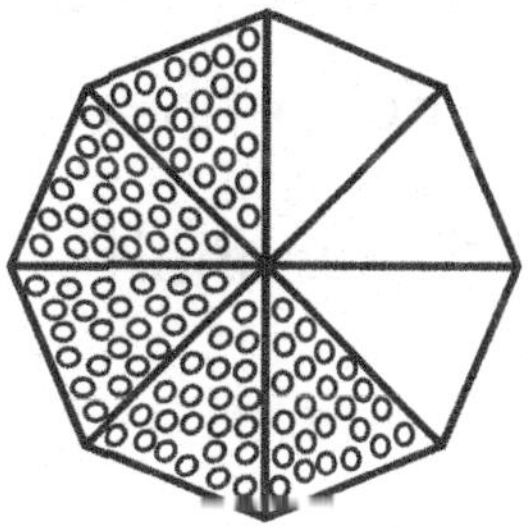

(C) 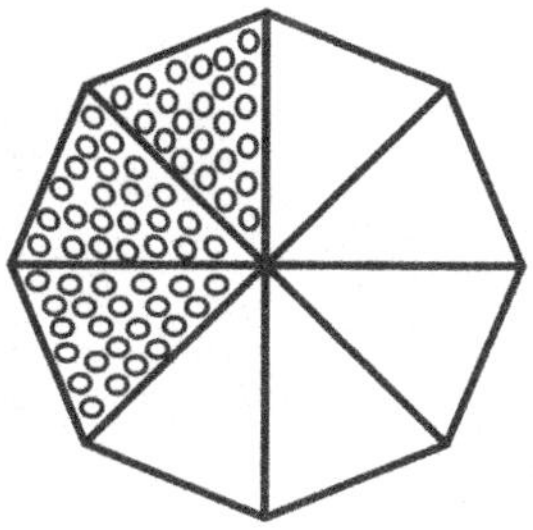

(D) 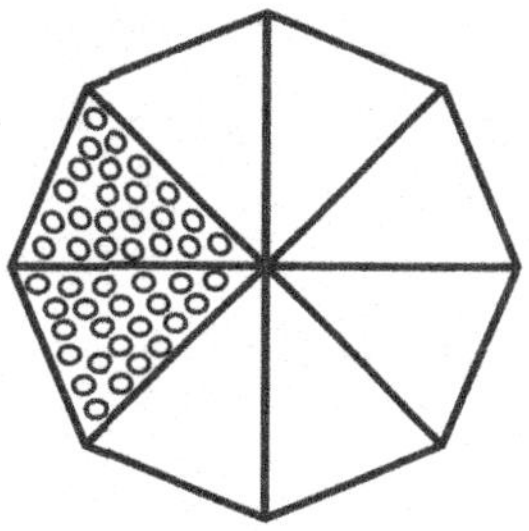

26. Weight of new born baby is measured in ?

(A) Cups (B) Ounces (C) Gallons (D) Pints

27. Tanya has 6 weeks of winter break. How many days was her winter break ?

(A) 7

(B) 30

(C) 12

(D) 42

28. Identify the closest temperature shown in the below figure ? (Approximately)

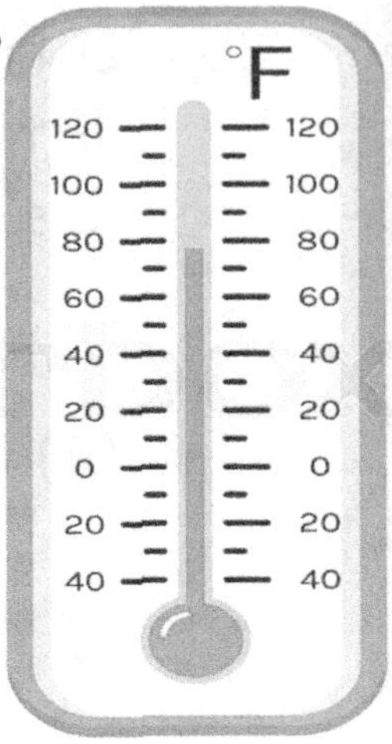

(A) 78^{0} Fahrenheit

(B) 150^{0} Fahrenheit

(C) 82^{0} Fahrenheit

(D) 13^{0} Fahrenheit

29. Identify the picture ?

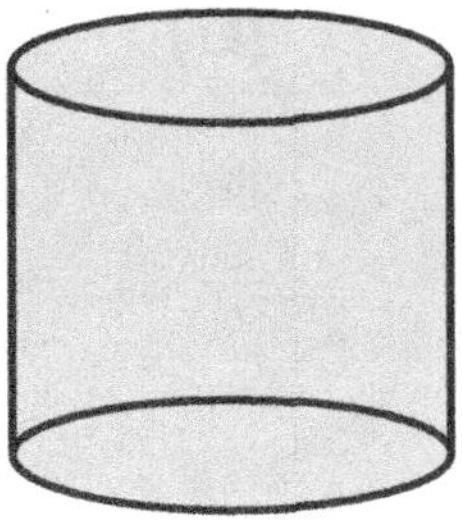

(A) Rectangle (B) Circle

(C) Cylinder (D) Sphere

30. Length of one tooth pick is equal to 1 inch. What is the closest length of the computer ?

(A) 1 inch (B) 8 inches

(C) 2 inches (D) 4 inches

31. Dan ties three Roses in a bunch to place them in the vases to decorate his sisters party tables. He has 18 roses in all. Which of the below shows the vases and Roses ?

(A) 

(B) 

(C) 

(D) 

32. Olivia is wondering which of the below is showing the time of 9.30 PM ?

(A) 

(B) 

(C) 

(D) 

33. Tall shooters basket ball club participated in various tournaments this year Below graph represents the number of games won across various months. What is missing on the graph ?

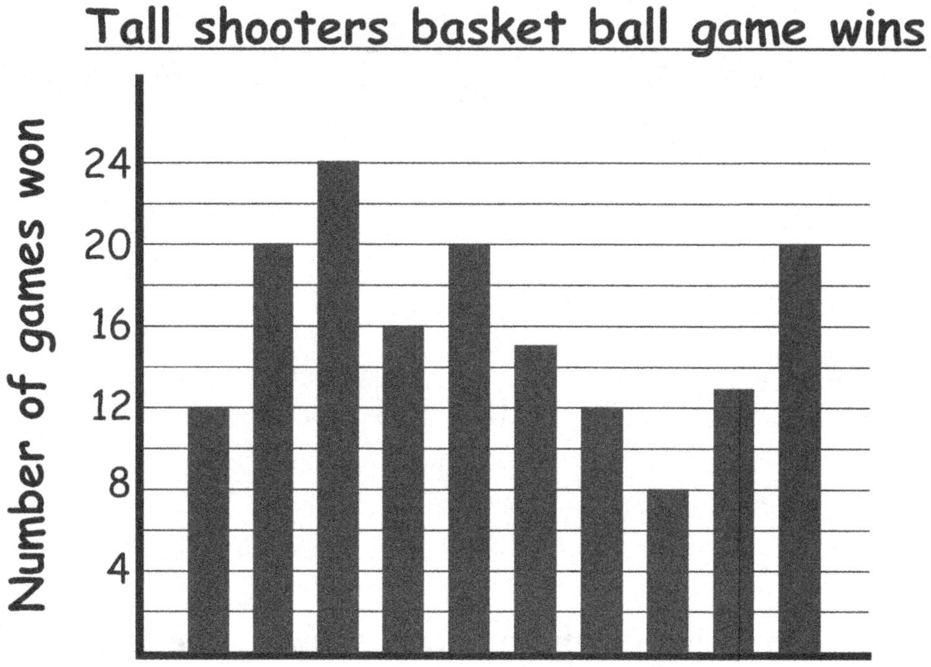

(A) Various months of participation

(B) Number of games participated

(C) Number of kids in the club

(D) Number of games lost

34. What is the perimeter to the distance around this figure in inches ? (Approximately using the ruler , assume all sides are of same length)

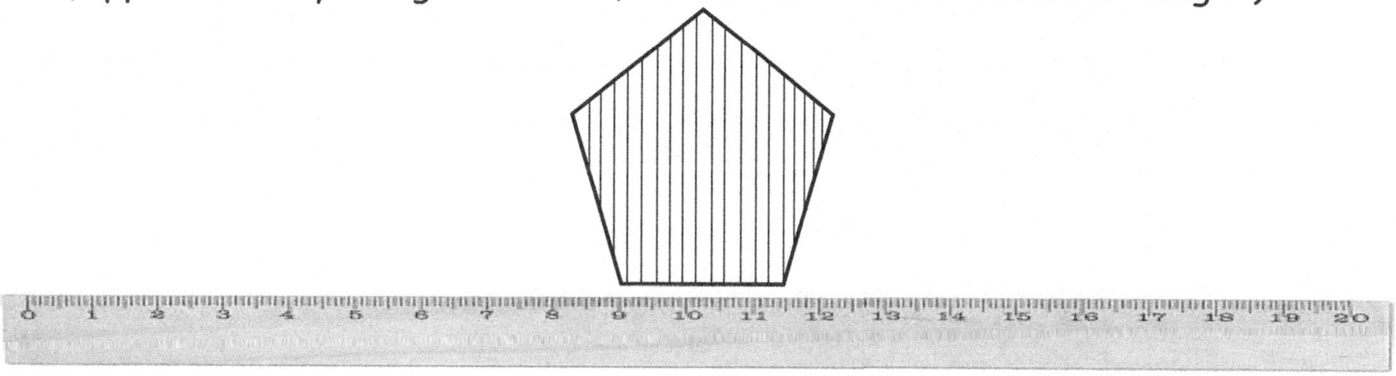

(A) 13 inches (B) 11.5 inches (C) 15 inches (D) 12.5 inches

35. Below data shows the number of bracelets made by Tara this quarter ?

Month	January - 2019	February - 2019	March - 2019
Number Repaired	35	21	42

She prepares the below chart representing the above data

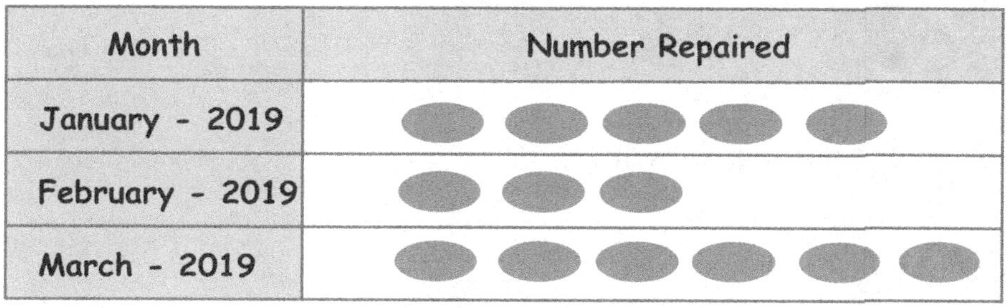

The symbol  represents the value of ____________ ?

(A) 2 (B) 7 (C) 5 (D) 4

36. Tony has

His saving this week from his weekly allowance is

How much money did he have in all ?

(A) $ 9.50

(B) $ 9.05

(C) $ 9.53

(D) $ 9.505

37. Lily collects below flowers from her flower garden to make various size flower bouquets. She wants to find out if she has enough flowers to make the bouquet orders ready for the day.

Which bar graph correctly shows the number of each kind of flowers that were collected today ?

(A)

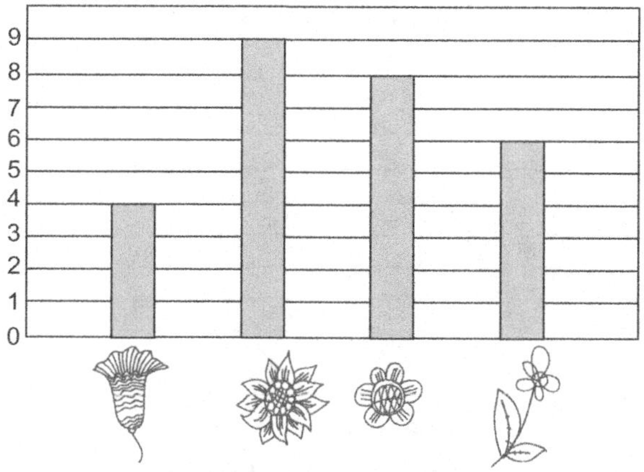

(B)

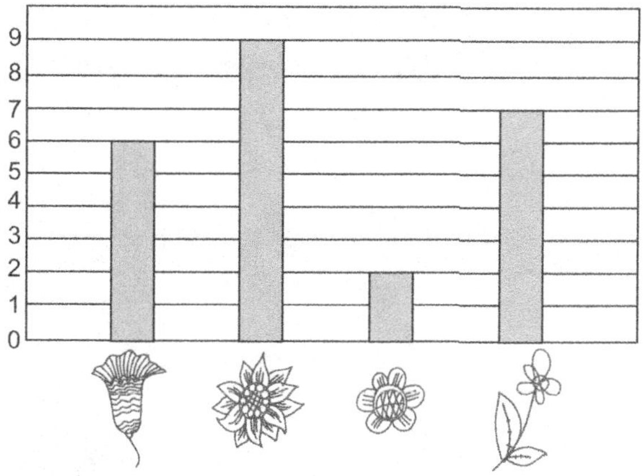

(C)

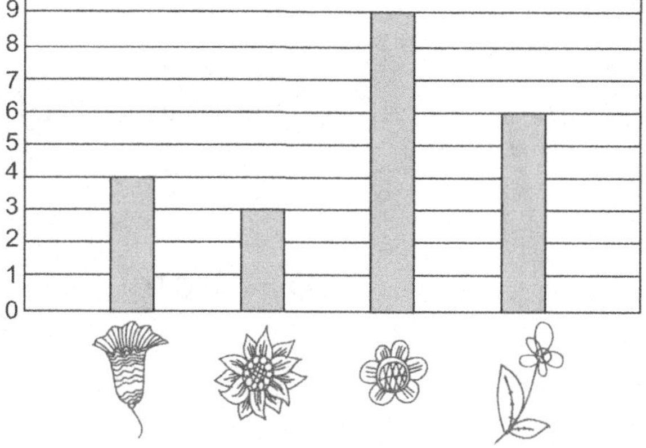

(D)

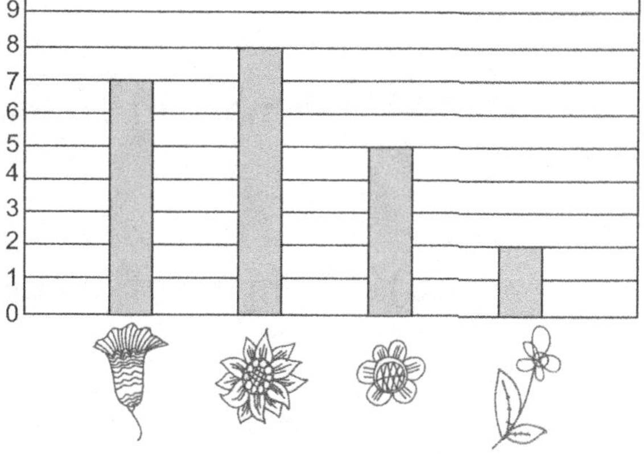

38. Jack picks an item from the box given below.
What are his chances of picking a tomato from the box ?

(A) Certain

(B) Likely, but not certain

(C) Unlikely, but not impossible

(D) Impossible

39. Grade 3 of Liberty school collects stamps across various months of this year Below graph shows the months horizontally.
What is missing vertically ?

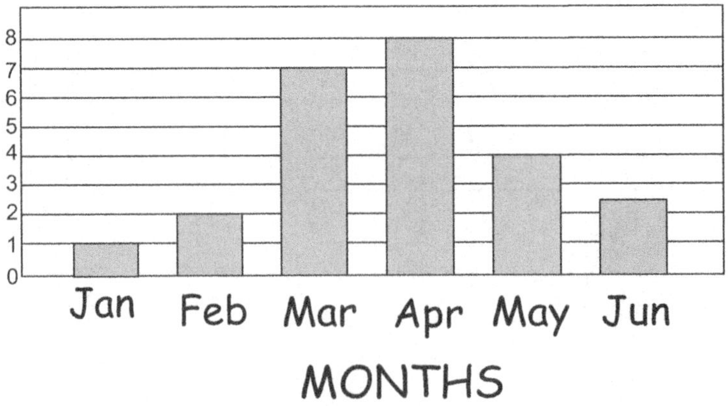

(A) Number of Stamps collected

(B) Number of Months

(C) Number of Weeks

(D) Total

40. Alice colored small squares in the shape of an alphabet from the below grid. If □ is equal to one square centimeters. What is the area of the alphabet "A" ?

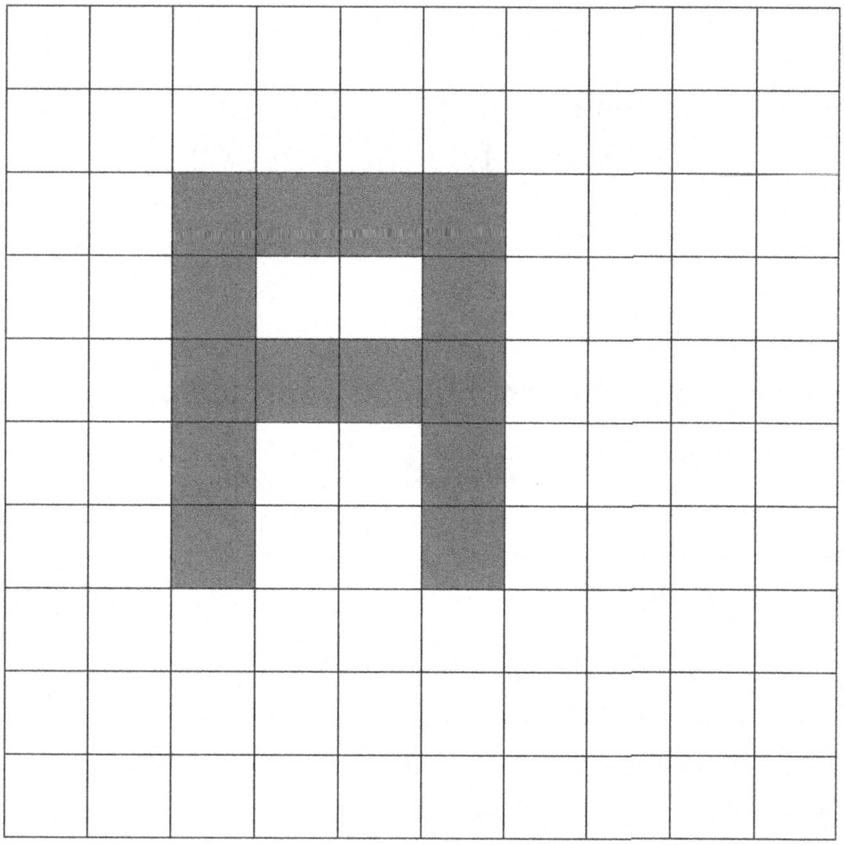

(A) 16 (B) 14 (C) 20 (D) 15

41. Find the next number in the series

45 , 54 , 63 , 72 , 81 ,.....?

(A) 92 (B) 91 (C) 90 (D) 93

42. Which number can be placed on the line to make the number sentence true ?

45 X ? = 90

(A) 45 (B) 2 (C) 1 (D) 3

43. Using the skip counting rule what are the next two numbers in the series ?

413 , 415 , 417 , 419 , 421 , 423 ,... ?

(A) 424 , 427 (B) 425 , 527 (C) 426 , 428 (D) 425 , 427

44. The table below gives the cost of Blue Berries at Tony's farm.

Blue Berries sale

Number of Blue Berries	Total Cost
1	$ 0.11
2	$ 0.22
3	$ 0.33
4	$ 0.44
5	$ 0.55
6	$?

Based on the pattern in the table, what will be the cost for six pounds of Blue Berries ?

(A) $ 0.60 (B) $ 0.77 (C) $ 0.56 (D) $ 0.66

45. Mona has scarfs ,shirts and trousers in her trunk. She counts them after sorting. Which chart below shows the correct count ?

(A)

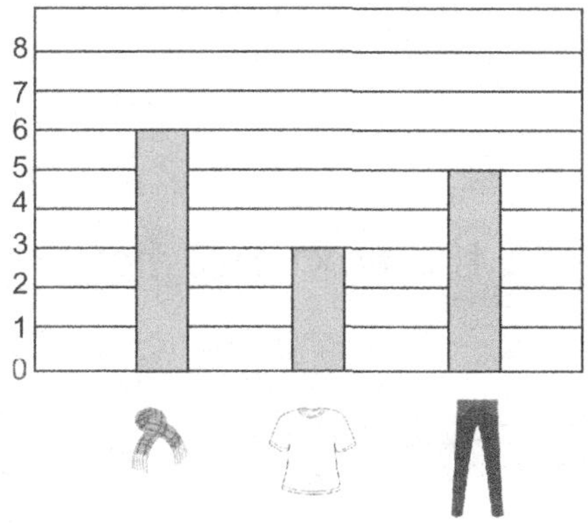

(B)

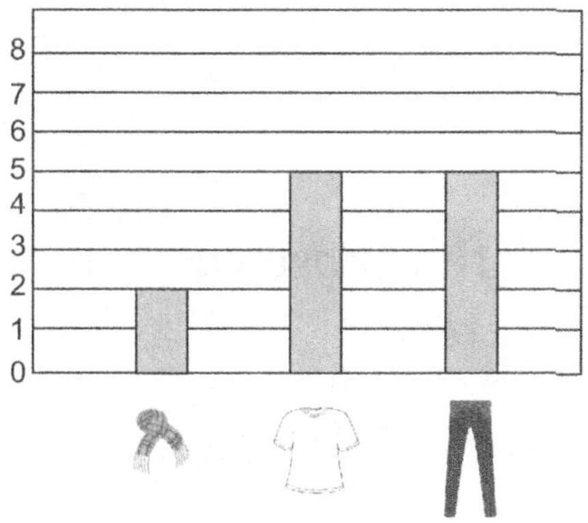

(C)

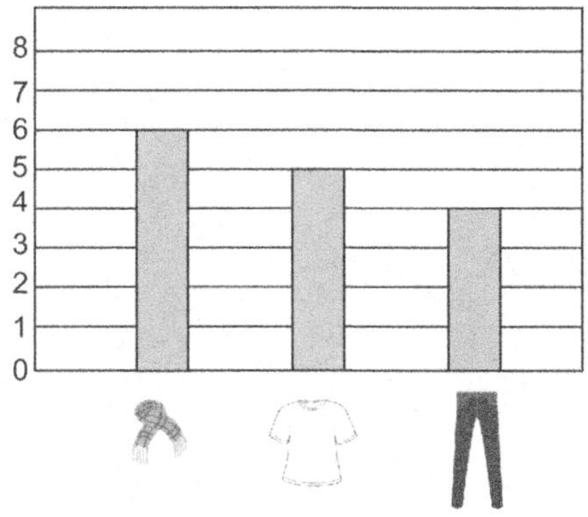

(D)

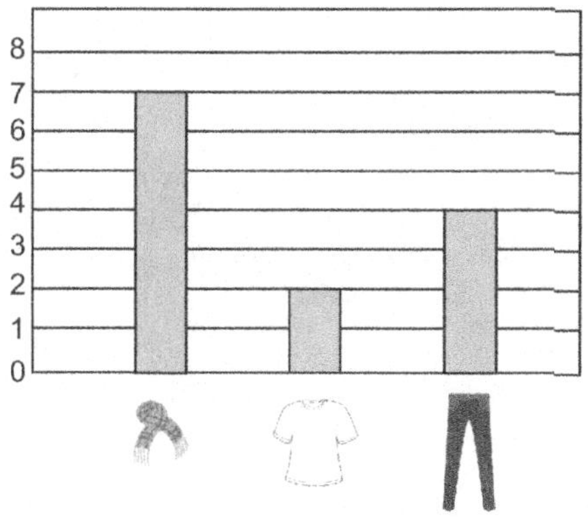

46. Esha uses the math fact $\boxed{9 \times 4 = 36}$ to solve a math problem. Which of the below could be the problem she is trying to solve ?

(A) $\square \div 4 = 9$

(B) $\square \times 9 = 4$

(C) $\square + \square = 36$

(D) $36 - \square = 9$

47.

...

If the above pattern continues, what are the next two figures in the pattern ?

(A) 

(B) 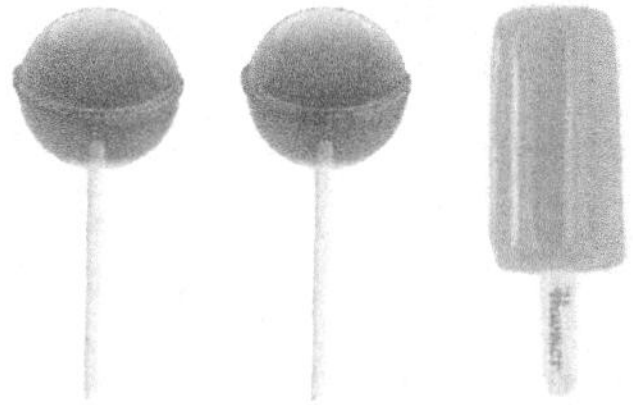

(C) 

(D) 

48. Kate planted her plants in various rows in her garden.
How many plants were planted in row five ?

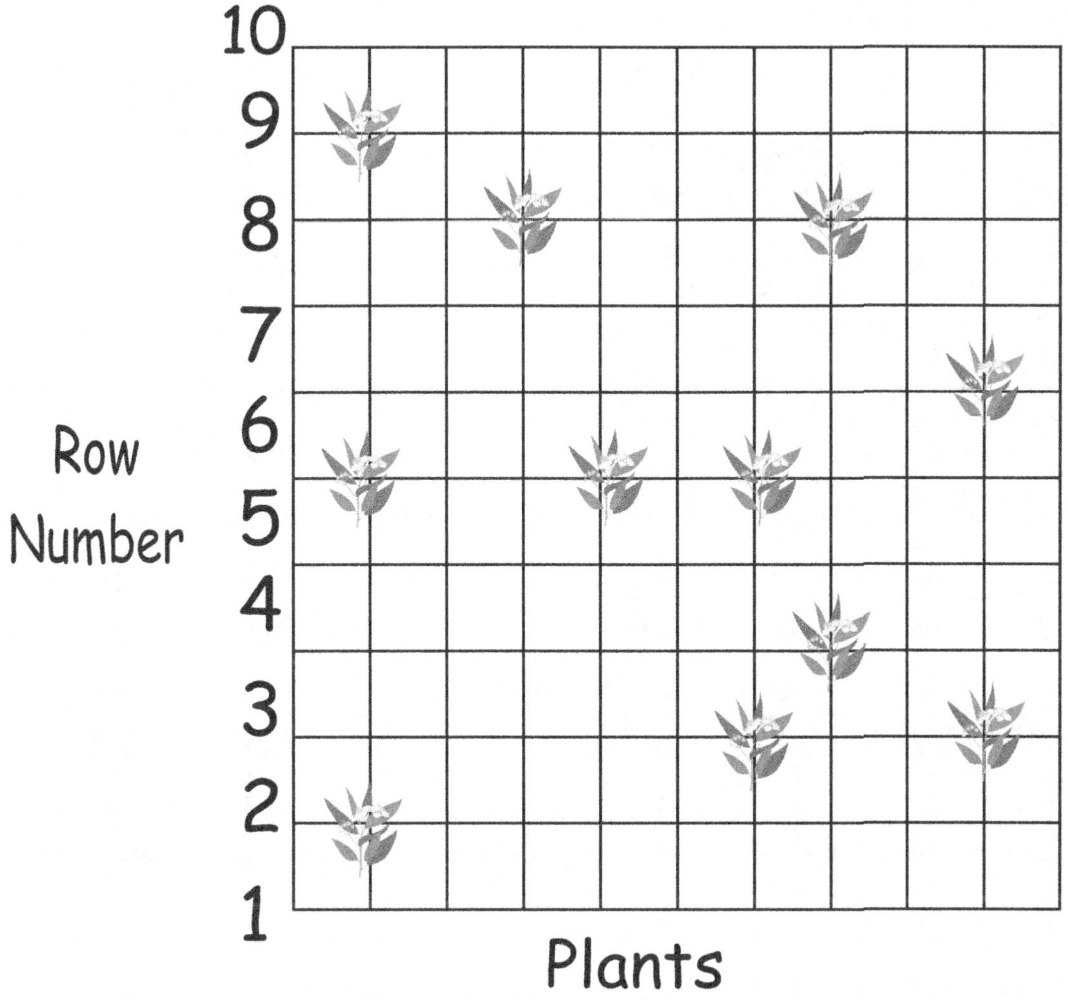

(A) 7

(B) 3

(C) 6

(D) 1

49. Grade 3 teachers from Ace elementary school took a survey in their respective classes on their favourite subject math. Below graph represents the number of votes received.

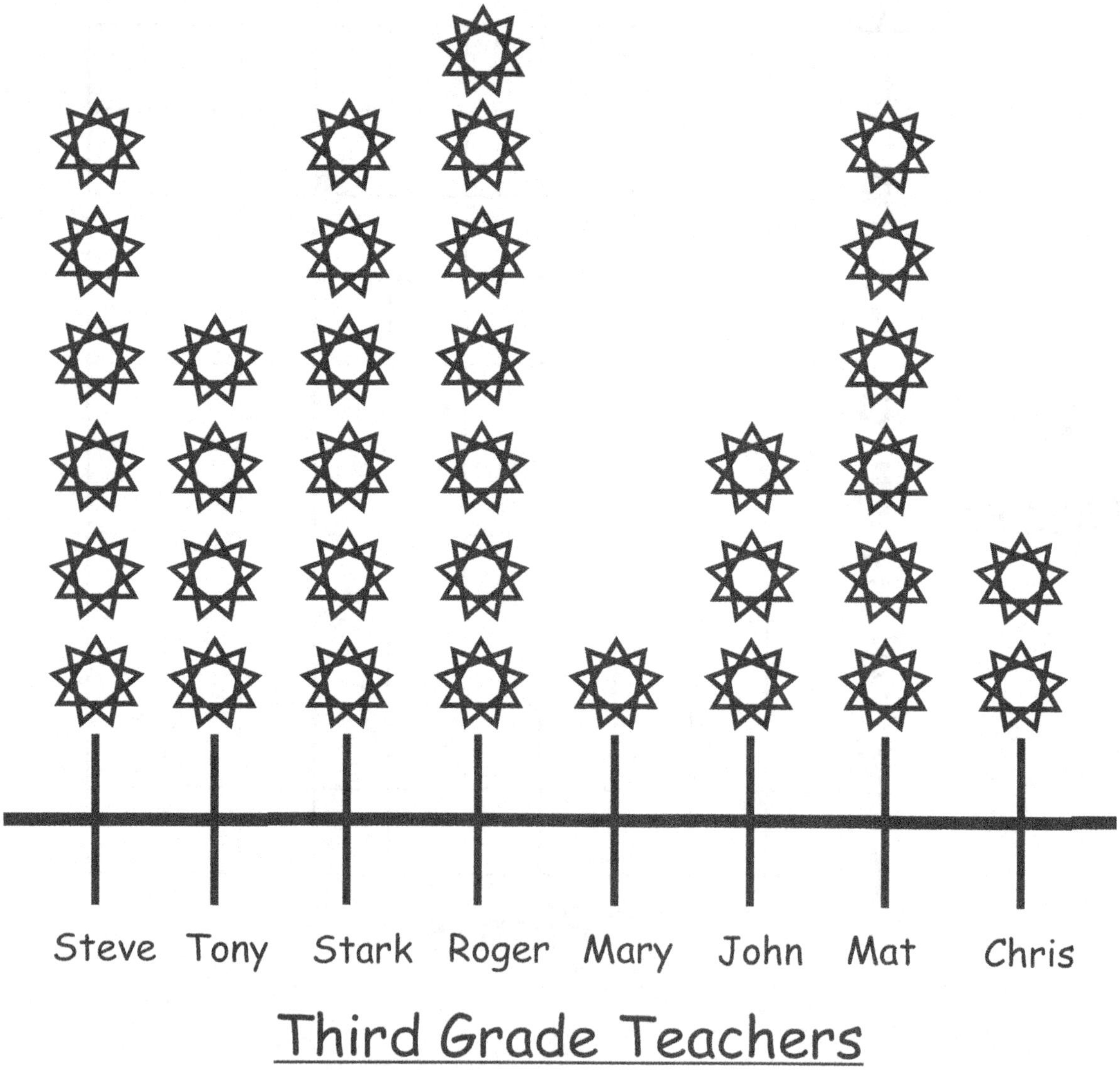

Which class students like the math most ?

(A) Mary

(B) Steve

(C) Stark

(D) Roger

50. Victor wants to decorate his Christmas tree. He sorts Christmas ornaments in group of 4 from his collection of twenty - eight ornaments to decorate Christmas tree. Which of his below grouping shows as a multiplication fact ?

(A)

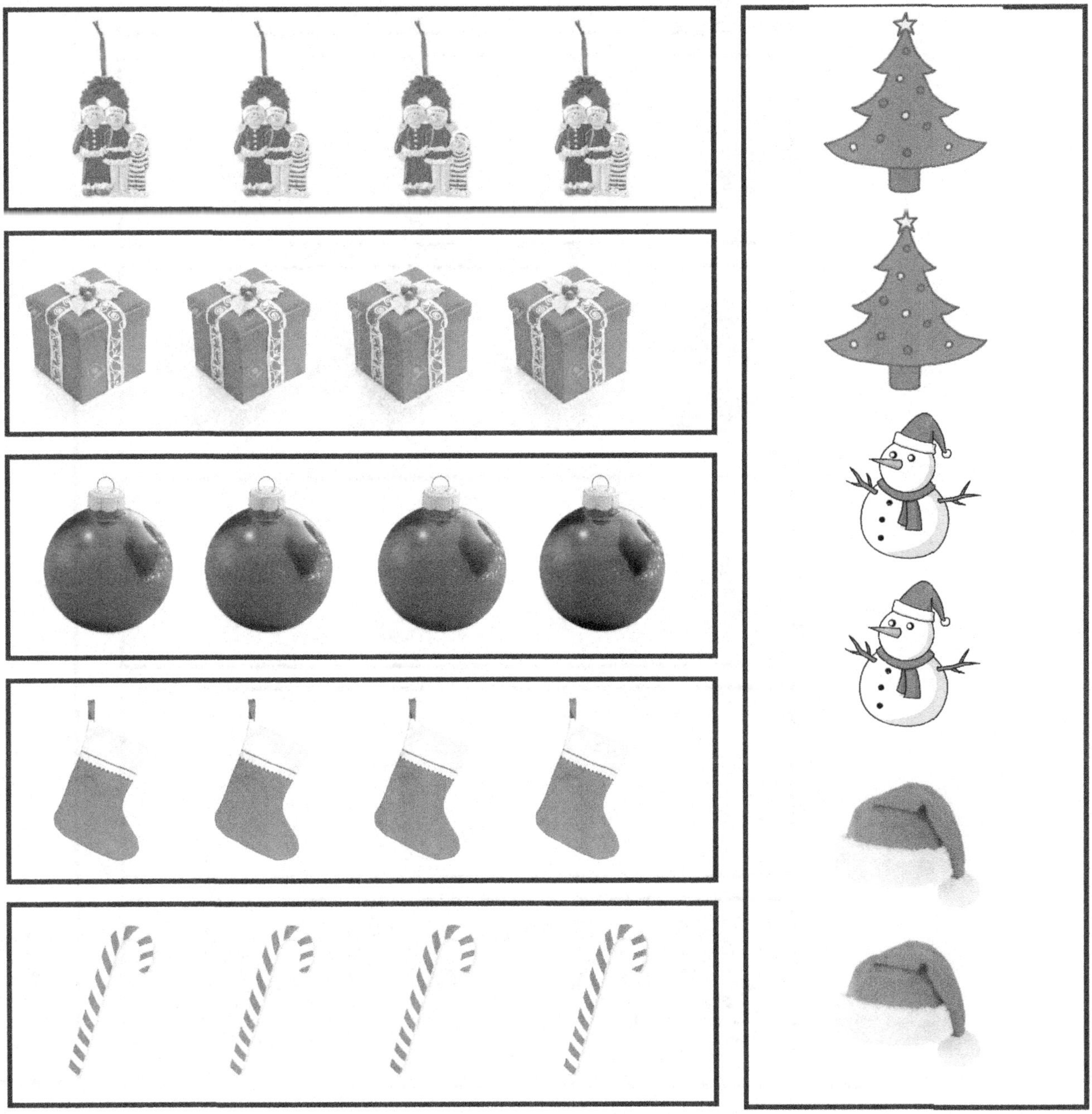

(B)

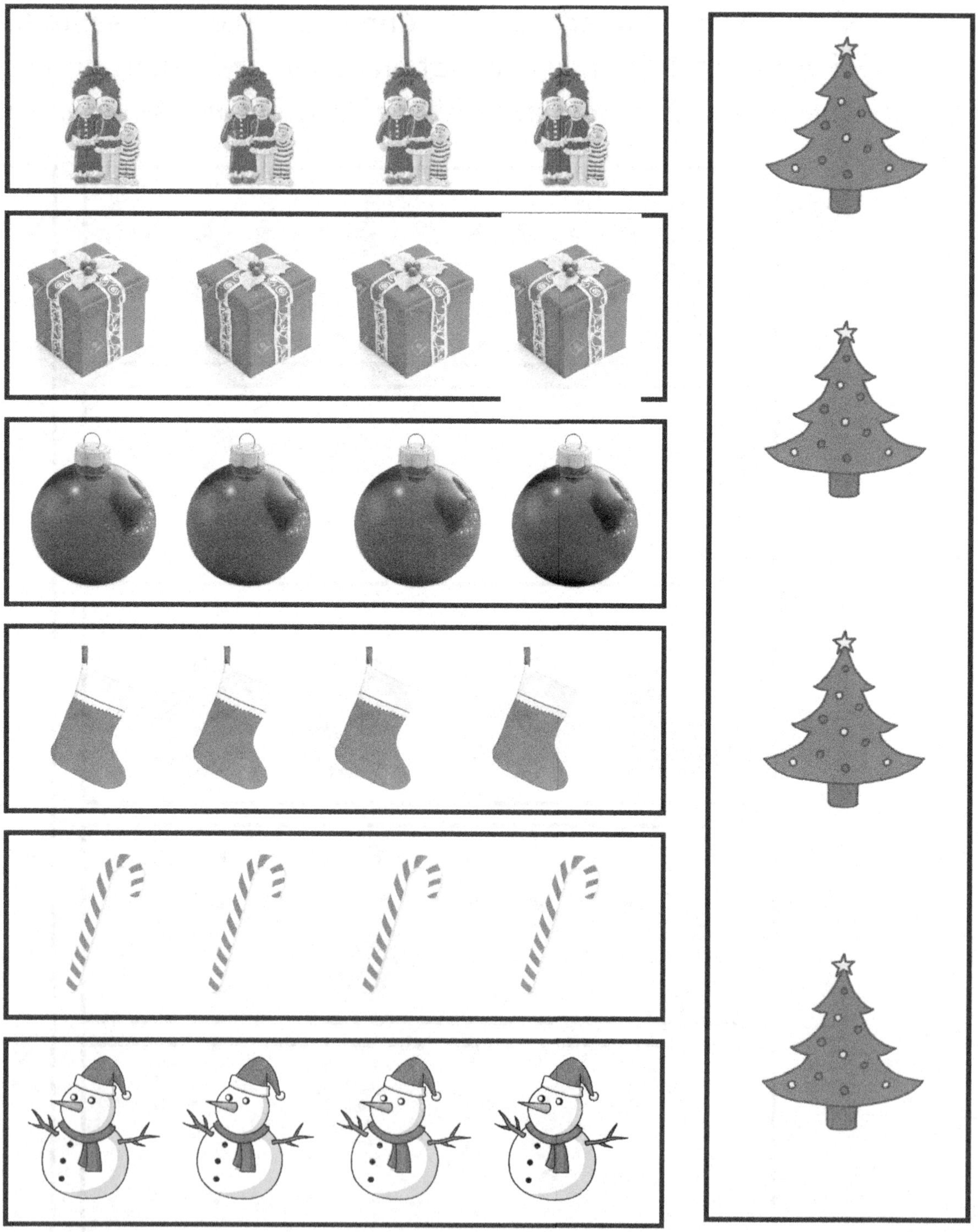

(C)

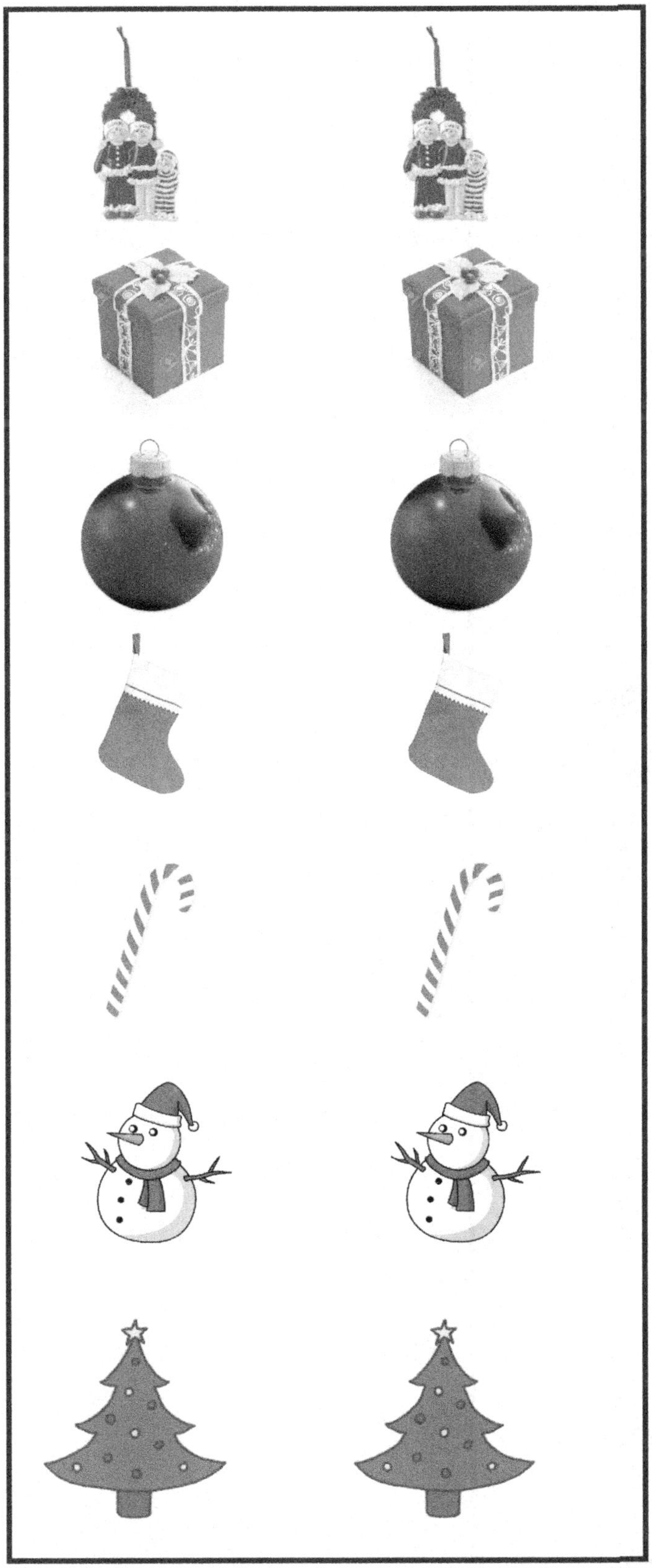

(D)

Grade 3
SOL
Practice Test - 6

1. Riya bought a house for $199,003 last week. What is 199,003 rounded to the nearest hundred thousand ?

   (A) 200,000

   (B) 200,003

   (C) 199,000

   (D) 199,103

2. Which of the below shows "Eight nine thousand, fifty-five" in standard form ?

   (A) 809,505

   (B) 89,550

   (C) 809,055

   (D) 89,055

3. Lucy and her team raised $4,559,700 for a non-profit organization to serve the needy. What is the value of 4 in 4,559,700 ?

   (A) 4000

   (B) 4,000,000

   (C) 40,000

   (D) 400,000

4. Which of the below statement is true ?

(A) 7,470 > 7,479 (B) 7,470 = 7,000

(C) 7,470 > 7,490 (D) 7,470 > 7,460

5. Tyler biked 10,589 miles this year. What is the value of eight in 10,589 ?

(A) 80,000 (B) 800

(C) 80 (D) 8,000

6. The below figure represents one whole part

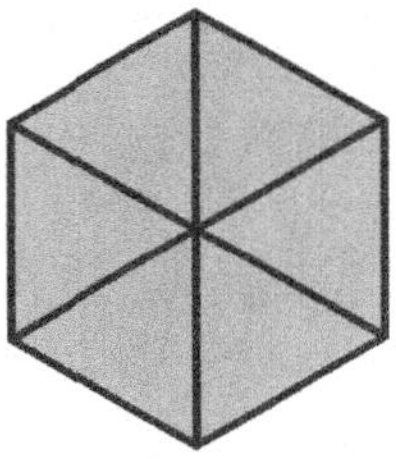

What fraction number is represented by the below ?

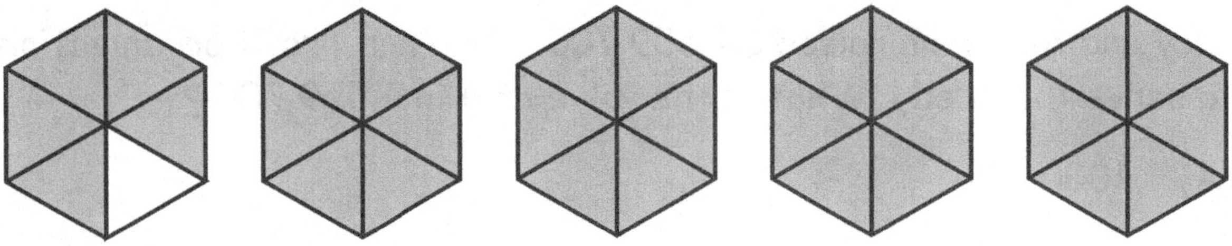

(A) $4\frac{5}{6}$ (B) $4\frac{2}{6}$

(C) $4\frac{3}{5}$ (D) $4\frac{1}{6}$

7. The figure below is shaded to represent the number 1.

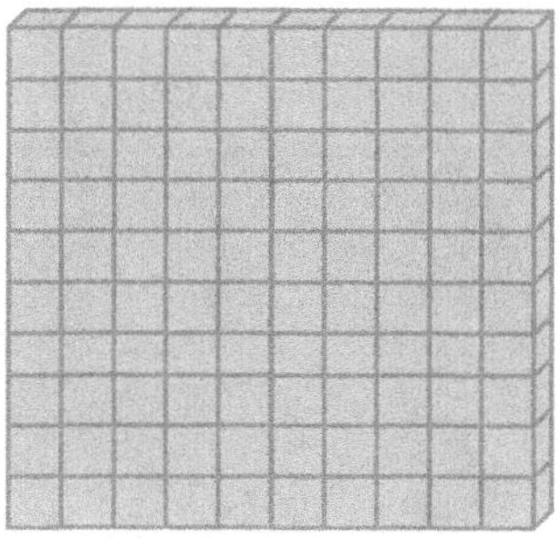

The below model represents which number ?

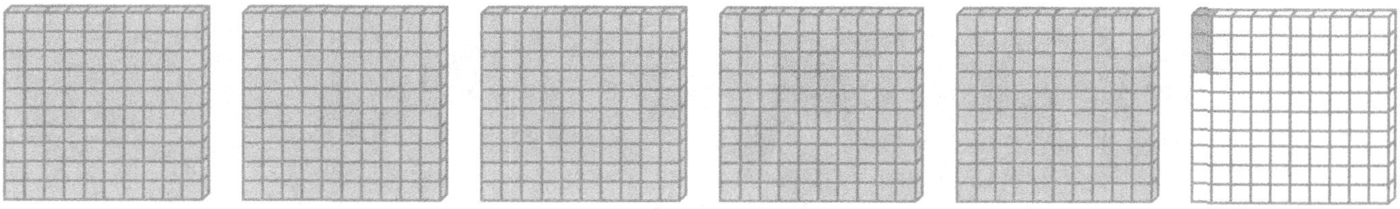

(A) 4.03 (B) 5.30 (C) 5.33 (D) 5.03

8. This model is shaded to show a fraction of a whole.

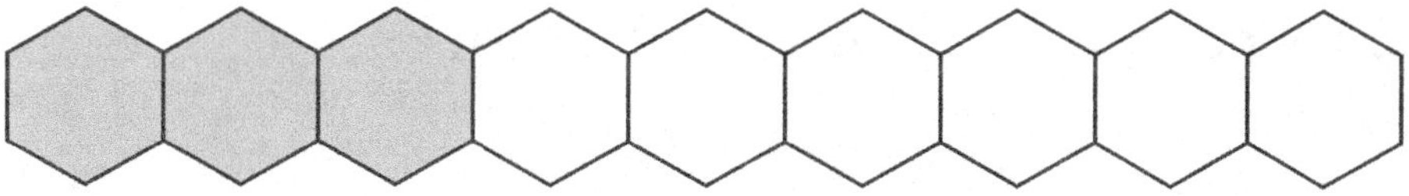

Which of the following is shaded to show a fraction with a value EQUAL the one above?

(A) 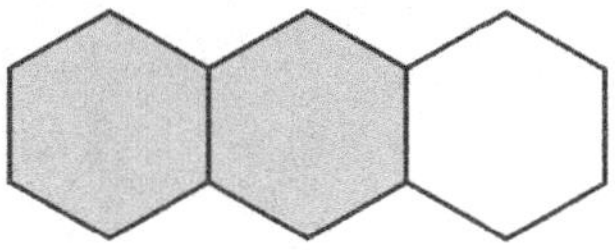

(B) 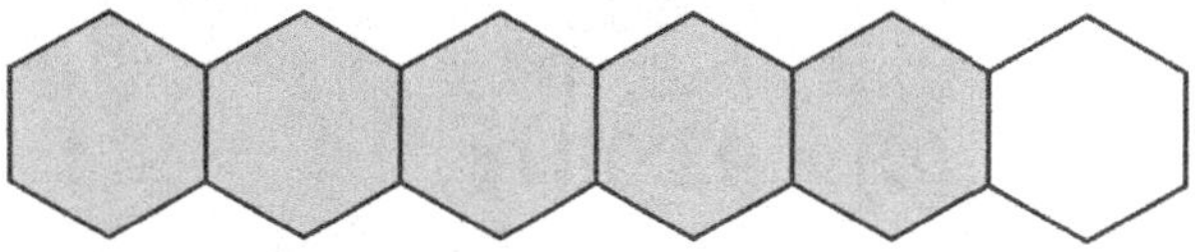

(C) 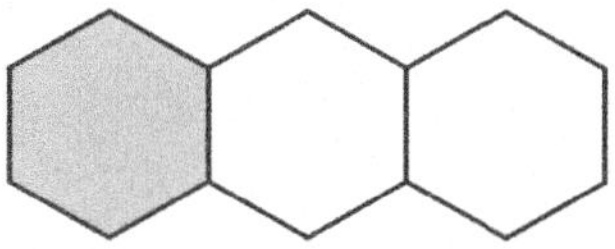

(D) 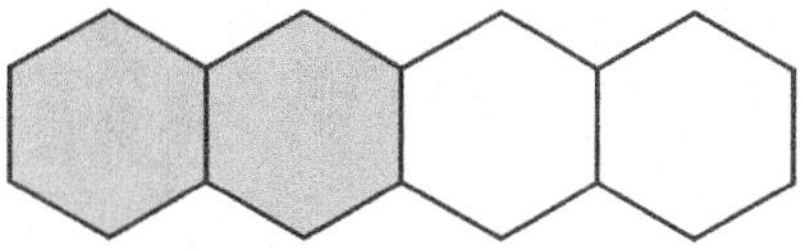

9. Which picture box below has least number of Arrows ?

(A)

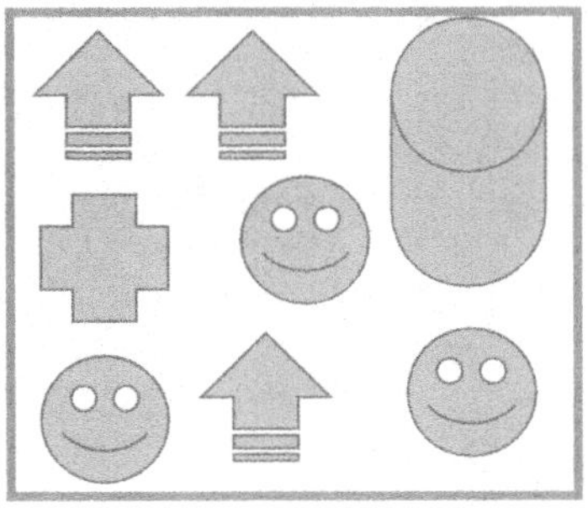

(B)

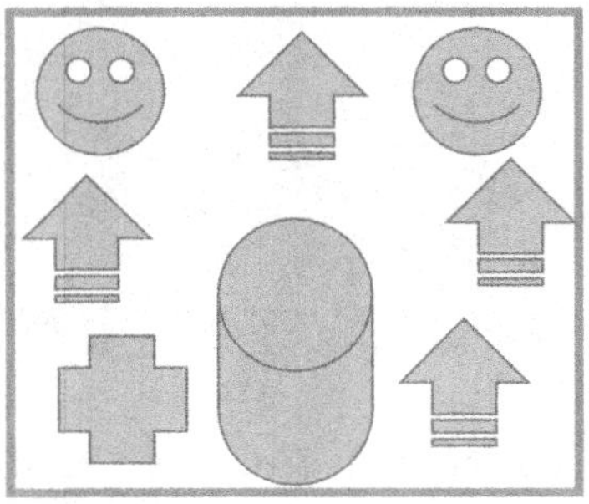

(C)

(D)

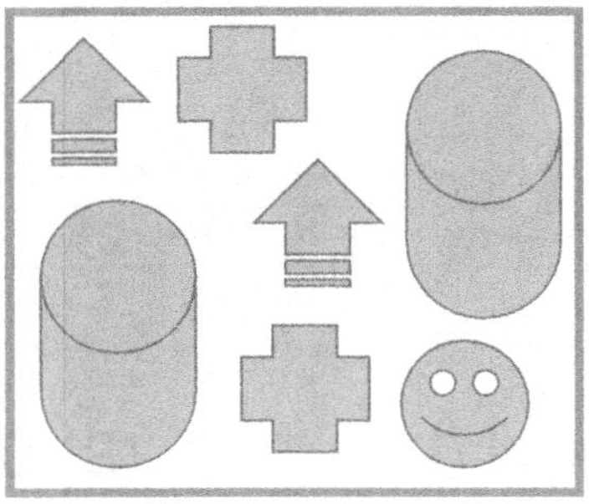

10. What is 8,885 rounded to the nearest hundred ?

(A) 8,900

(B) 8,800

(C) 9,000

(D) 8,880

11. Dan has 5 children. If he gets two gifts for each of his children, How many gifts did he gave them in all ?

(A) 54

(B) 25

(C) 10

(D) 300

12. This  represents a value of 1.

Which model represents the value of 1 in 777,210 ?

(A) 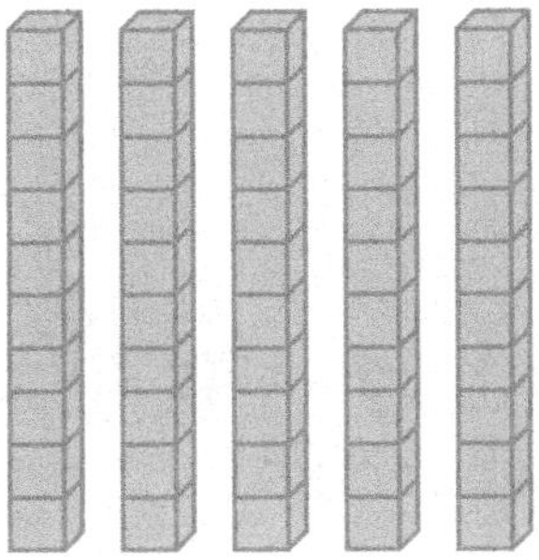

(B) 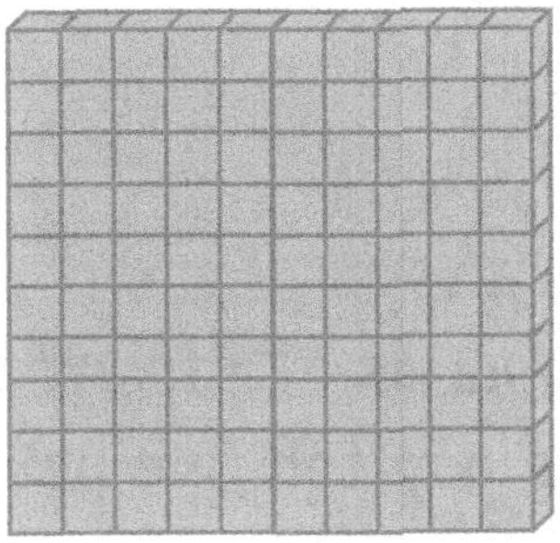

(C) 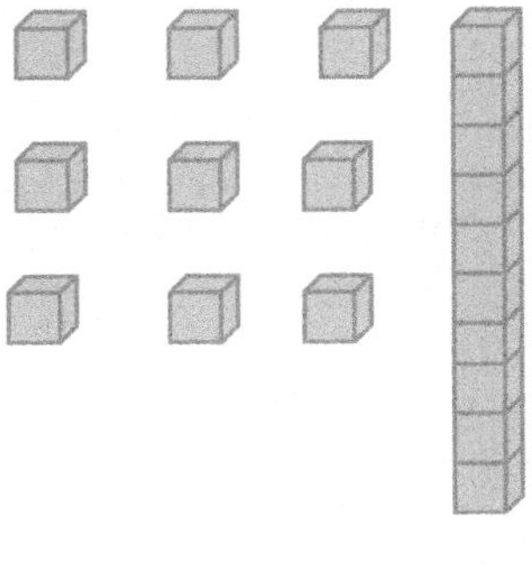

(D) 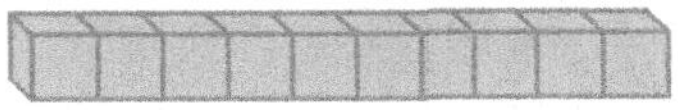

13. What fractional part of tennis balls are there from the below group ?

(A) $\frac{2}{3}$ (B) $\frac{1}{2}$ (C) $\frac{1}{6}$ (D) $\frac{6}{1}$

14. Which of the below statement is true ?

(A) 404 > 414

(B) 225 < 299

(C) 838 > 939

(D) 712 < 090

15. Jack gets five pounds of strawberries. If each pound is six dollars how much did he pay for five pounds of strawberries ?

(A) 30

(B) 10

(C) 25

(D) 11

16. Evaluate $2 \times 5 = \boxed{?}$

(A) 3

(B) 6

(C) 10

(D) 9

17. Zara has 17 pens and 22 books

17	22

How many less pens does she have than books ?

(A) 11

(B) 12

(C) 10

(D) 5

18. Evaluate $4441 + 2202 = \boxed{?}$

(A) 6,503

(B) 6,643

(C) 6,603

(D) 6,043

19. This model is shaded to show 1 whole.

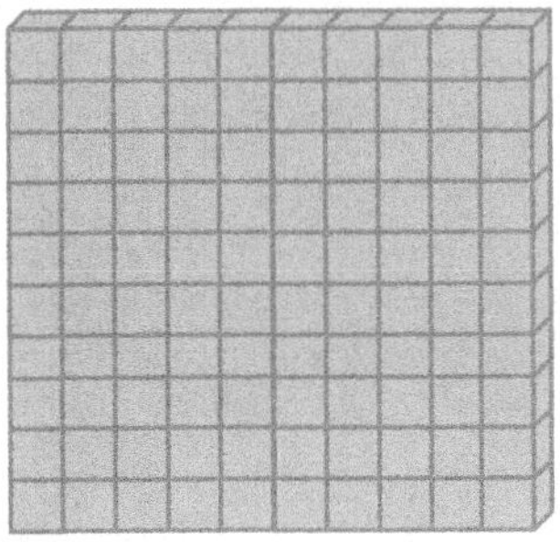

What decimal is shown by the model below ?

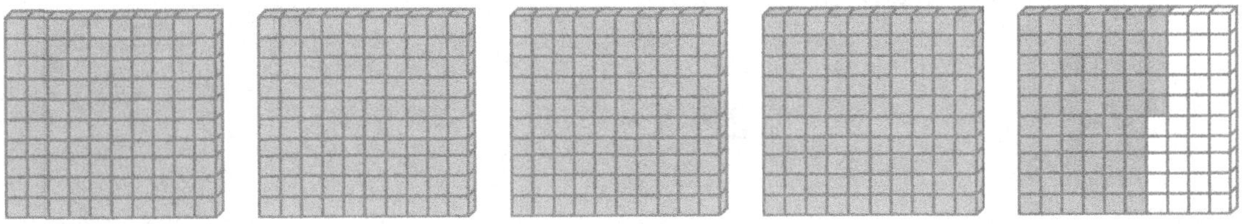

(A) 5.69 (B) 4.55 (C) 4.65 (D) 4.66

20. Evaluate

$$\begin{array}{r} 6\,.\,62 \\ +5\,.\,07 \\ \hline \\ \hline \end{array}$$

(A) 8.78

(B) 11.69

(C) 12.3

(D) 25.8

21. This model is shaded to show one whole.

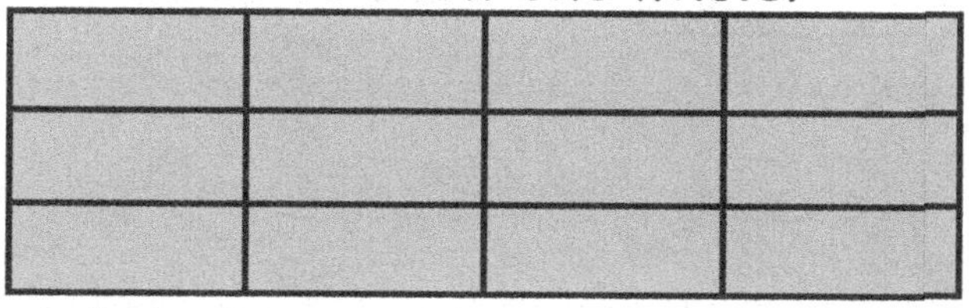

The two models below are each shaded to show a fraction

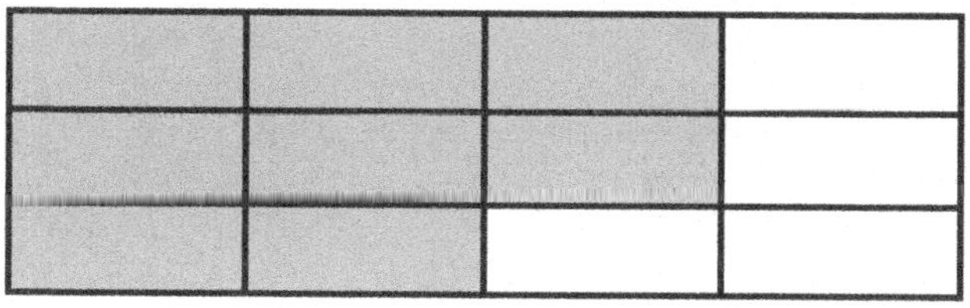

Model 1

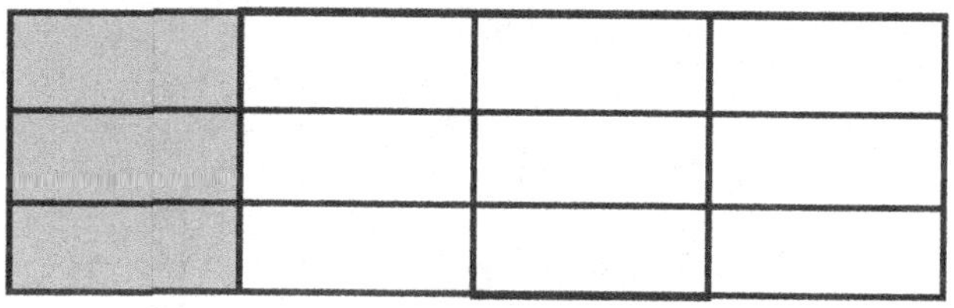

Model 2

One fraction is subtracted from the other. Which model is shaded to show the difference ?

(A) $\frac{5}{8}$ (B) $\frac{3}{8}$ (C) $\frac{11}{12}$ (D) $\frac{5}{12}$

22. Evaluate

$4 \times 9 = \boxed{?}$

(A) 81

(B) 36

(C) 01

(D) 90

23. Evaluate $6 \times 3 = \boxed{?}$

(A) 12

(B) 10

(C) 9

(D) 18

24. Evaluate $8.2 + 0.73 = \boxed{?}$

(A) 8.93

(B) 8.96

(C) 8.27

(D) 7.19

25. Which of the figures below appears to have a line of symmetry ?

(A) 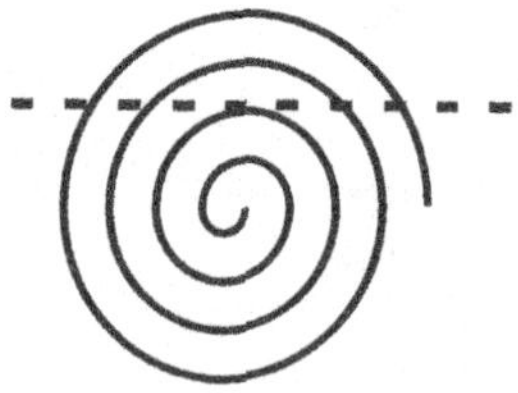

(B) 

(C) 

(D) 

26. Bella makes 88 glasses of fruit punch. She sells 44 glasses of fruit punch to her guests. What fractional part of the fruit punch is left ?

(A) $\frac{1}{2}$ (B) $\frac{3}{4}$

(C) $1\frac{1}{4}$ (D) $\frac{1}{4}$

27. Dimensions of the book is measured in which units ?

(A) Kilometers

(B) Gallons

(C) Pounds

(D) Centimeters

28. Siya solves one puzzle in 30 seconds. Time taken to solve 4 puzzles in minutes is ?

(A) 120

(B) 2

(C) 20

(D) 10

29. Identify the closest temperature shown in the below figure ? (Approximately)

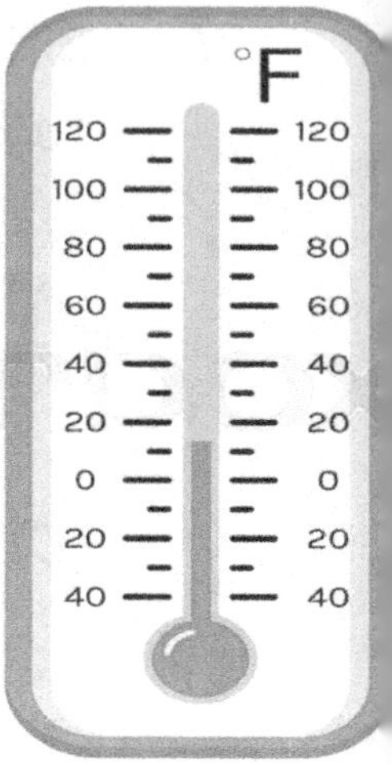

(A) 63^0 Fahrenheit

(B) 150^0 Fahrenheit

(C) 26^0 Fahrenheit

(D) 13^0 Fahrenheit

30. What is the closest to the length of tree ?

(A) 19 cm

(B) 22 cm

(C) 15 cm

(D) 11 cm

31. Which of the below picture represents a right angle ?

(A) 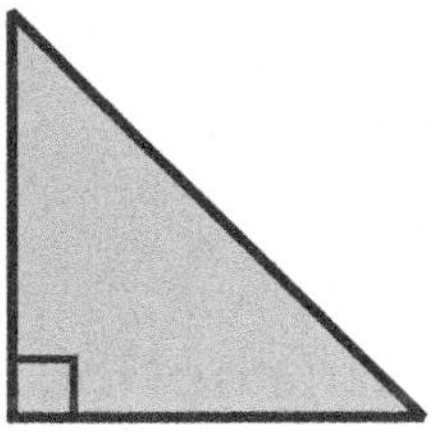

(B) 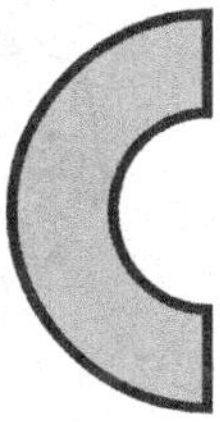

(C) 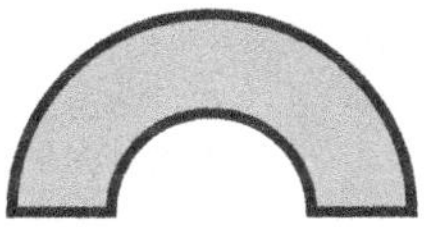

(D) 

32. Olivia is wondering which of the below is showing the time of 7.15 PM ?

(A) 

(B) 

(C) 

(D) 

33. This model is shaded to show one whole.

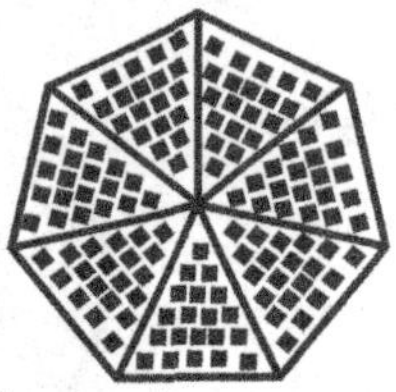

The two models below are each shaded to show a fraction

One fraction is subtracted from the other. Which model is shaded to show the difference ?

(A) $\frac{8}{7}$ (B) $\frac{1}{2}$ (C) $\frac{1}{7}$ (D) $\frac{4}{7}$

34. What is the perimeter to the distance around this figure in cm ? (picture not drawn to scale)

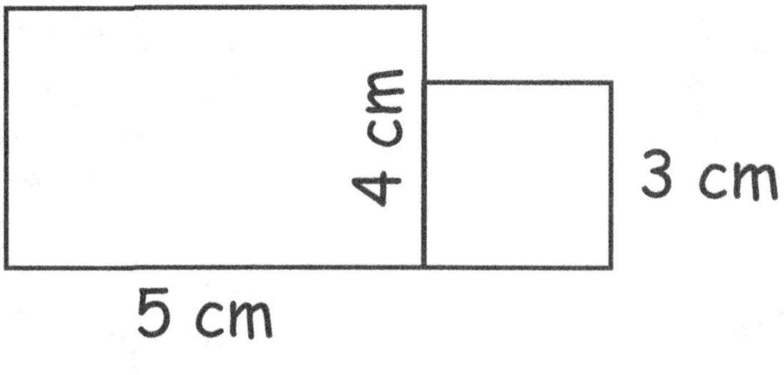

(A) 24 cm (B) 20 cm

(C) 27 cm (D) 12 cm

35. Ben has

He spent this week ?

How much money did he have now ?

(A) $ 8.50

(B) $ 4.35

(C) $ 5.46

(D) $ 5.25

36. Below data shows the number of cars repaired in Zack's garage this quarter ?

Month	May - 2019	June - 2019	July - 2019
Number Repaired	40	25	35

He prepares the below chart representing the above data

Month	Number Repaired
May - 2019	
June - 2019	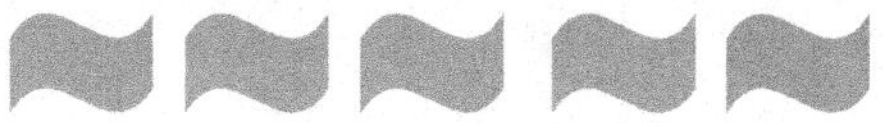
July - 2019	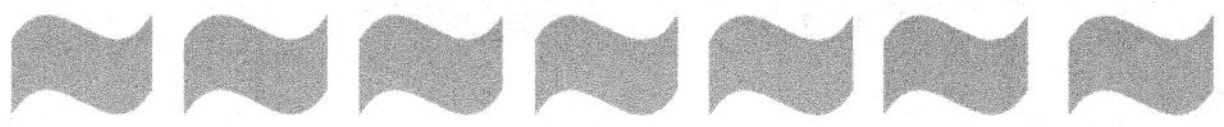

The symbol 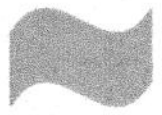 represents the value of ____________ ?

(A) 2

(B) 5

(C) 3

(D) 4

37. These are leaves Sandy collected for school.

Which bar graph correctly shows the number of each kind of leaf ?

(A)

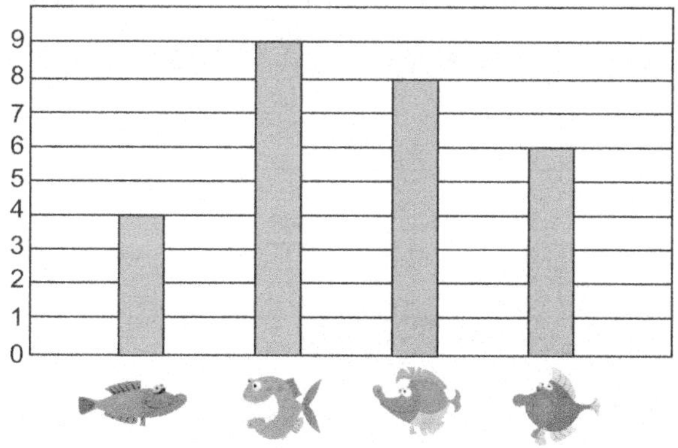

(B)

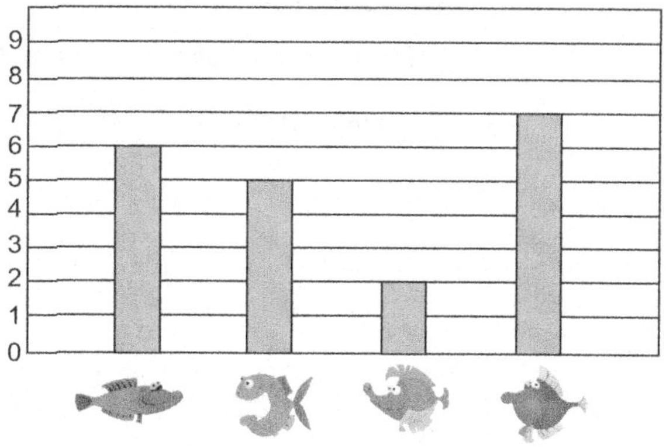

(C)

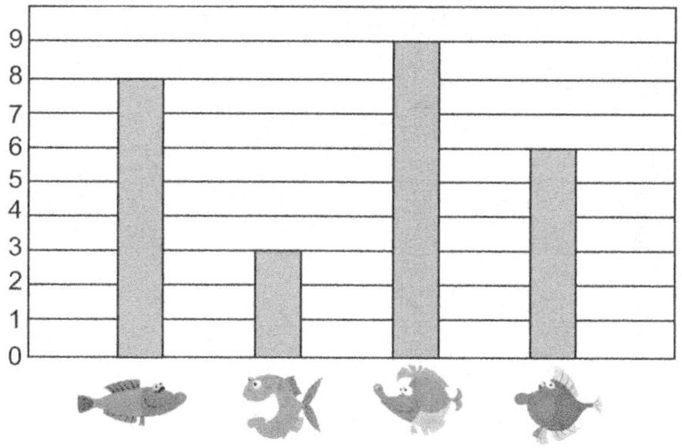

(C)

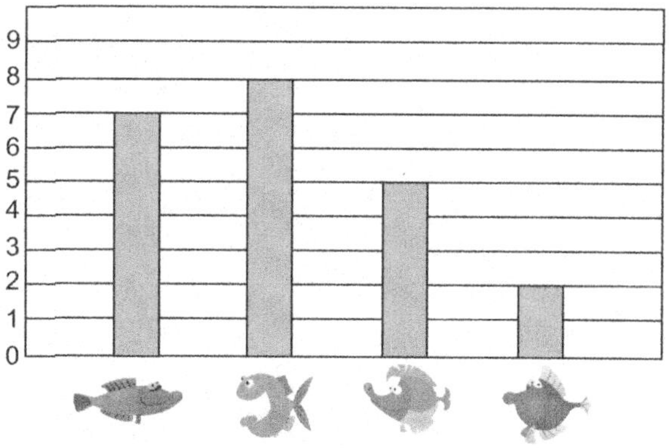

38. Below numbers and alphabets are written on individual cards and are placed in a jar after carefully placing each of them in envelops. Tracy picks a card from the jar.
What is the chance of an alphabet in the envelop ?

KK	10	RP	1	GV	GK	2	5	MV

(A) Certain
(B) Likely, but not certain
(C) Unlikely, but not impossible
(D) Impossible

39. Alice colored small squares in the shape of an alphabet from the below grid. If □ is equal to one square centimeters. What is the area of the alphabet "P" ?

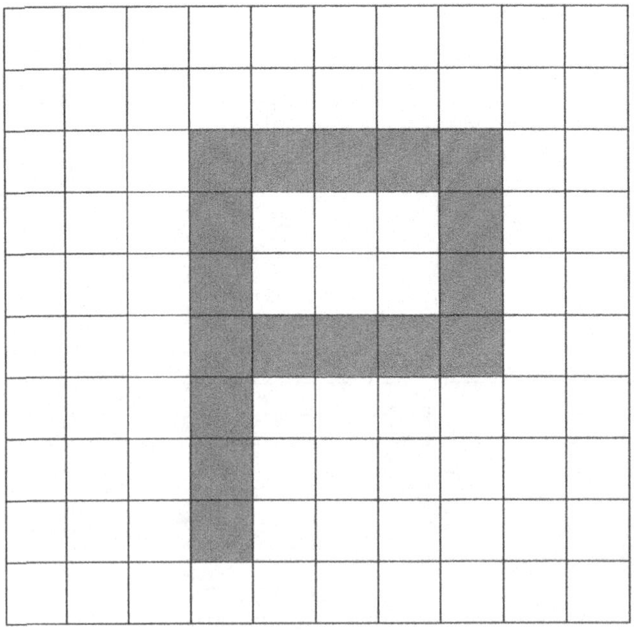

(A) 16 (B) 14 (C) 20 (D) 17

40. Which of the below statement is true ?

(A) 107 is greater than 173

(B) 912 is less than 670

(C) 1838 is equal to 1668

(D) 2075 is less than 19999

41. Find the next number in the series ?

5 , 10 , 15 , 20 , 25,........?

(A) 35 (B) 40 (C) 30 (D) 45

42. Which number can be placed on the line to make the number sentence true ?

25 X [?] = 75

(A) 0 (B) 1 (C) 2 (D) 3

43. Using the skip counting rule what is the next number in the series ?

600 , 606 , 612 , 618 ,.... ?

(A) 624 (B) 625 (C) 626 (D) 630

44. The table below shows the costs of brownies in David high school.

Brownie sale price

Number of Art Pieces	Total Cost
1	$ 0.60
2	$ 1.20
3	$ 1.80
4	$ 2.40
5	$ 3.00
6	$?

Based on the pattern in the table, what will be the total cost for six brownies ?

(A) $ 3.30 (B) $ 3.60 (C) $ 4.10 (D) $ 4.60

45. A square has ...... ?

(A) All sides are not equal

(B) Two pairs of non parallel sides

(C) Opposite sides are parallel

(D) All of the above

46. Brian has two types of dice in his math club. He is sorting the dice in the following two groups. He placed dice of the same shape and sides into a group.

Group six side dice

Group eight side dice

Which shape should Brian include in Group eight side dice ?

(A)  (B) 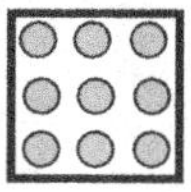 (C)  (D) None

47. Evaluate $9 \times \square = 81$

Which of the below fact can be used to complete the given number sentence ?

(A) $81 + 9 = 9$

(B) $81 \times 9 = 18$

(C) $81 - 9 = 9$

(D) $81 \div 9 = 9$

48. Crazy Art club created various art pieces for the last seven weeks. Below graph represents the number of art works created across the weeks. What is missing on the graph ?

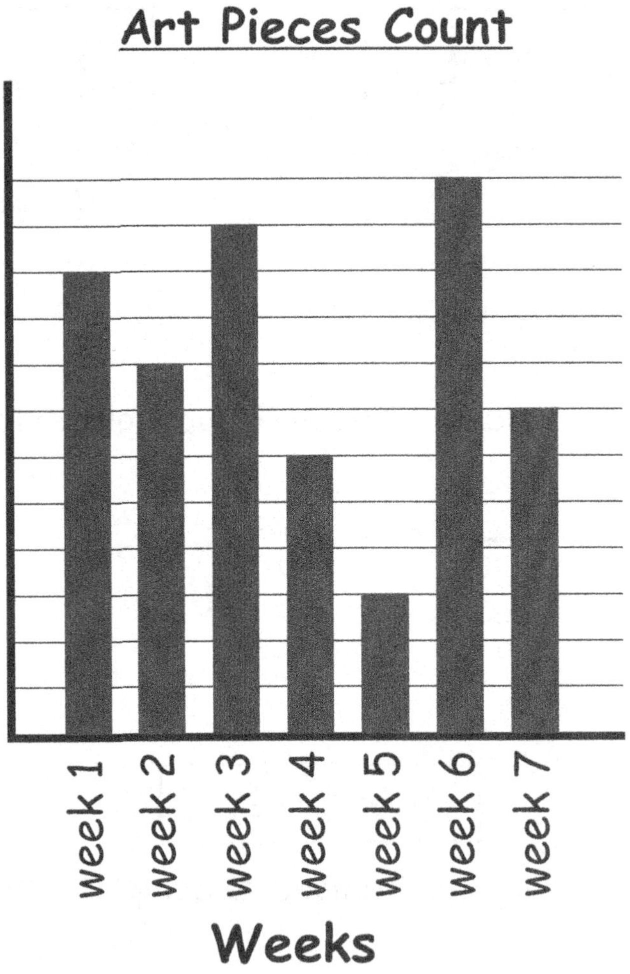

(A) Number of Art pieces created

(B) Average number of Art pieces

(C) Number of awards won

(D) Number of people in Art club

49. Olivia bakes 25 cupcakes to sell in her bakery. She displace them in the groups of five. Which of her below grouping shows as a multiple fact ?

(A)

(B)

(C)

(D)

50. The number of baskets made by tiny shots basket ball team members on memorial day tournament are plotted below.

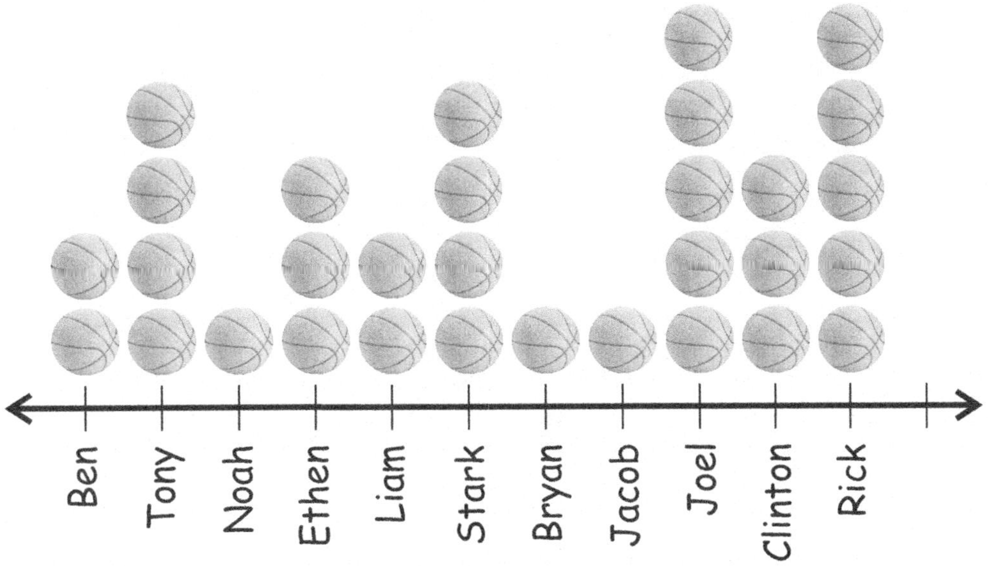

How many baskets were made in total by the team ?

(A) 5

(B) 22

(C) 31

(D) 35

GRADE 3
SOL
Practice Test - 4
Answer Keys

Answer Key Test - 4

1. D
2. C
3. A
4. D
5. C
6. C
7. D
8. C
9. A
10. B
11. D
12. D
13. A
14. C
15. A

16. C

17. B

18. A

19. D

20. B

21. A

22. C

23. D

24. B

25. D

26. B

27. C

28. A

29. D

30. C

31. A

32. D

33. C

34. A

35. C

36. A

37. D

38. A

39. B

40. C

41. B

42. A

43. D

44. C

45. A

46. D

47. D

48. C

49. A

50. C

GRADE 3
SOL
Practice Test - 5
Answer Keys

Answer Key Test - 5

1. D
2. C
3. A
4. C
5. B
6. D
7. B
8. D
9. A
10. C
11. A
12. B
13. B
14. D
15. C

16. A

17. D

18. C

19. A

20. D

21. A

22. D

23. C

24. A

25. C

26. B

27. D

28. A

29. C

30. D

31. B

32. C

33. A

34. D

35. B

36. C

37. A

38. D

39. A

40. B

41. C

42. B

43. D

44. D

45. C

46. A

47. C

48. B

49. D

50. B

GRADE 3
SOL
Practice Test - 6
Answer Keys

Answer Key Test - 6

1. A
2. D
3. B
4. D
5. C
6. A
7. D
8. C
9. D
10. A
11. C
12. D
13. A
14. B
15. A

16. C

17. D

18. B

19. C

20. B

21. D

22. B

23. D

24. A

25. C

26. A

27. D

28. B

29. D

30. A

31. A

32. C

33. D

34. A

35. C

36. B

37. A

38. B

39. D

40. D

41. C

42. D

43. A

44. B

45. C

46. C

47. D

48. A

49. A

50. C

GRADE 3
SOL
Score Calculation

Score calculation

If you get this many times correct :	Then your converted scale scor is :
0	000
1	217
2	249
3	268
4	282
5	293
6	303
7	311
8	319
9	325
10	331
11	337
12	343
13	348
14	353
15	357
16	362
17	366
18	371
19	375

20	379
21	383
22	387
23	391
24	395
25	399
26	403
27	407
28	411
29	415
30	419
31	423
32	427
33	432
34	436
35	441
36	445
37	450
38	456
39	461

40	467
41	473
42	480
43	487
44	496
45	505
46	516
47	531
48	550
49	582
50	600

Made in United States
North Haven, CT
06 November 2026

21791727R00072